LARS KINTER

THE HYGGE
ATTITUDE

•• •• ••

SWITCH YOUR LIFE
WITH THE DANISH
MINDFULNESS
OF HAPPINESS AND
COZINESS

Table Of Contents

INTRODUCTION TO HYGGE THE DANISH PHILOSOPHY OF HAPPINESS

Aren't we all looking for happiness? As it seems to me, the Danes know exactly how to find it! After all, Denmark is regularly in first place for the happiest countries in the world. How do you find such happiness for yourself, and is there really a secret recipe for such luck? And it looks like the answer may be "yes." Many Danes swear the secret is *hygge*, so I'm going to explain what it's all about and why the Danish art of luck can conquer the world.

1. WHAT IS HYGGE?

One could say that the Danish concept of *"hygge"* is a pleasant atmosphere or a feeling of coziness. The adjective *"Hyggelig"* literally means "pleasant," "cozy," and "nice," and originally "careful" or "thoughtful." Over time, however, it has been assigned the meaning "spreading welfare."

As we can see, *hygge* can have different connotations, all of which are positive. *Hygge* is the art of creating a pleasant atmosphere and the ability to find happiness in small pleasures, according to Meik Wiking, author of the book *Hygge — An Attitude That Will Just Make You Happy*.

If we asked a Danish person what the *hygge* philosophy is and why it enjoys so much popularity, they would certainly just start laughing, because *hygge* is a natural way of life for all Danes. It's so natural for them that most are stunned that *hygge* has become a kind of export hit in the whole world. After all, to be happy, it's enough to be yourself, to enjoy life, and to do what you feel good about. The author of the book

Hygge: The Big Luck, Marie Tourell Søderberg, says it lies in the little things. She emphasizes that *hygge* can be found everywhere. You just have to know what to look for.

A good comparison is the idea of holidays — the family is together, everywhere smells like delicious food, and the whole apartment is lovingly decorated... you can relax. That's pure *hygge*! The secret of the Danes is to keep that feeling alive not just at Christmas but throughout the year.

Hygge does not have to be expensive either — quite the contrary. Following the principle of "less is more," *hygge* remembers the little pleasures in life. Deceleration, just relaxing or enjoying a little mobile- and computer-free time — *hygge* can find in many everyday things. Awareness of the here and now, enjoying the moment and doing one after the other — these are values that are valued by *hygge*.

In this sense, typical *hygge* activities are easy to integrate into everyday life — just drink a hot cup of tea while reading a book, take a warm bubble bath, cook or bake, play board games with friends, or hone your craft skills in knitting or painting classes. It's just a healthy and enjoyable lifestyle.

THE RECIPE FOR HYGGE:
LUCK, NO MATTER WHERE YOU ARE!

The Danish philosophy of happiness is now also very popular with people across the world. The Collins dictionary even identifies the term "*hygge*" as one of the most popular words of the year, and designers reflect the concept of warmth and coziness with natural materials, while the Internet is full of fashion, blogs, and tutorials designed to help implement "*hygge*."

The word "coziness" denotes a special way of being together. It depends on where you are, who you are with, and what you are doing. What you consider *hygge* in practice can be very different from person to person. Some will say that they enjoy being by themselves, but they often have informal contact with family and friends. It will typically take place at home or in other types of intimate spaces. When you enjoy yourself, you often eat and drink together. On the whole, coziness is the expression of the simple joys in life. It does not necessarily take many things to create a cozy atmosphere; e.g., just a few candles or a bowl of Friday candy can make a difference.

The idea of coziness reflects some broader

values in Danish society. Coziness is closely associated with perceptions of equality. In nice times, inequality, power differences, or disagreements should not be too obvious. The importance of coziness in Denmark is probably also related to the Danish weather, where cold, dark, and wet months call for socializing within.

While most people consider coziness as something positive, coziness can also be introverted and exclusive. Few people enjoy themselves with someone they do not know, and it can be difficult to access the homes where the fun unfolds.

Is coziness, especially Danish? No, in practice, people enjoy themselves in other countries as well, but they do not necessarily attribute the same meaning as they do in Denmark. In other words, it is the *idea* of coziness, which has a special significance in Denmark, and which Danes often use to explain to foreigners what is Danish.

That's why *hygge* is important for the future:

Hygge is a social activity that can help create communities in everyday life. *Hygge* is part of many of the proposals that have been submitted to the Danish canon: a social gathering, food traditions, festive traditions, Danish design, song culture, book reading, seeing EM 92 together, and even going to the festival. In other words, hygiene means a great deal to many Danes' own perceptions of what is typical of Denmark and essentially of Danish culture.

At the present time, we are very absorbed in our work and the constant pursuit of a better future. We constantly set new goals, and face new challenges to make a career, make more money, buy new cell phones, cars, or fashionable clothes. With the development of modern technology and easy Internet access, we spend most of our free time reading news online and watching the lives and achievements of our friends, as well as intently watching the lives of celebrities we will never meet. Often (perhaps unconsciously), we compare ourselves with others, which only leads us deeper into the vortex of ambition and the pursuit of success.

Many people complain about the lack of time and the overwhelming fatigue caused by the demands of the modern world. Most of us do not have spare time on weekdays, because we are over-absorbed in daily obligations. Almost no one has time to sit down and meditate, read a book, or meet friends for coffee. Sometimes, we have to stay longer at work, do the grocery shopping, fill out paperwork, meet deadlines, pick up our kids from school or kindergarten, wash the car, and cook dinner. Once you've finally done housework, laundry, ironing, and so on, you quickly realize that another day has passed, and it's time to go to bed.

This is how weeks and months pass: we feel overwhelmed by everyday life and become more and more stressed. And although it seems our bank account has more money, it does not mean that we live better and that our lives look like we hoped. How do we get out of this situation? What do we have to do to be happy, even though the world around us is moving faster and faster?

2. HYGGE IN YOUR OWN HOME

As already mentioned, *hygge* describes much more a lifestyle than a certain style. Therefore, *hygge* is not limited to one location, but it is still very important in your own home. Your own home should be a place of security and a place to feel good. "*Hygge*" means to appreciate the small things and is therefore reflected in minimalist style elements. Perhaps you already know this style, where minimalism is by no means cold and clinically equable but rather represents a feeling of order, coziness, and reflection on the essentials. CANDLES, plants, cuddly blankets, and dresses as well as special keepsakes (e.g., framed photos of family and friends) embody this attitude and are therefore central decoration elements.

Likewise, the rooms themselves need in no way be designed around a clinical white. Although "*hygge*" refers serenity and the liberation of unnecessary ballast, it also focuses on home and individuality. Wallpaper is, therefore, a great way to decorate your *hygge*-style home to your liking. With wallpaper, you can easily create a connection to nature, whether with a delicate floral motif, photo wallpaper with trees, or wallpaper in a wood look. This can give your home a proper *hygge* atmosphere. In addition, wallpaper gives your walls more structure and thus provides a change.

When designing your home, you can also set up special places in your rooms for specific purposes. For example, is there anything better than a cozy reading corner? With larger rooms, you can easily use different wallpapers to separate certain areas or set highlights. Pay attention to your dining room, your bed, or the cozy sofa corner where friends and family can come together.

The Meaning of *Hygge*: Simply Explained

The word "*hygge*" actually comes from Norway. Later, it also appeared in Danish and became popular there because of how it describes a very specific way of life.

"*Hygge*" can be translated as "well-being." Furthermore, *hygge* describes a cozy atmosphere in which one feels comfortable and relaxed.

The term can be applied to all areas of life. Whether it's visiting family, meeting friends, or attending street parties — *hygge* is an integral part of

Danish culture. When setting up, the Danes pay attention to the *hyggelig* feeling, such as a cozy atmosphere with candlelight.

Hygge plays a big role, especially in winter. Then the Danes meet with friends or relatives for a cozy get-together, sitting for hours at the table and philosophizing about life.

If you travel to Denmark once, you will feel the same in the cafes, in the parks, or even at one of the numerous festivals.

If you also want a *hyggelig* home, you should paint your walls with warm colors, put on cuddly fabrics, and, above all, set up many candles.

Origin of *Hygge*

"*Hygge*" did not originate in the Danish language but rather in Norwegian, where the term means "well-being." The term only appeared in the Danish written language toward the end of the 18th century, and since then, the Danes have adopted the term. The good thing about "*hygge*" is that it can be integrated into all contexts. The Danes manage to transfer the *hygge* mood to all commonplace situations — and you'll certainly notice that during a holiday in Denmark!

3. HYGGE IN DAILY LIFE

"*Hygge*" is a core component of the Danish tradition. In essence, "*hygge*" is a cozy, warm atmosphere in which to enjoy the good of life with nice people. The warm light of candles is *hygge*. Friends and family are also part of the "*hygge*." And not to mention eating and drinking — for Danes, that preferably means sitting for several hours at the table and to deal with the big and small things in life together. Perhaps the Danish phenomenon of "*hygge*" explains why the Danes are often considered to be one of the happiest peoples in the world?

***Hygge* Season: *hygge* is available year-round**

When seeking *hygge*, it really does not matter when you come to Denmark and whether you prefer to travel to the North Sea or to the Baltic Sea, because *hygge* is an inherent part of Danish culture: you will find *hygge* in the decor of your holiday home or hotel ("Danish design" is the keyword), in the cozy little cafes, and, of course, in the form of culinary delights in the bakeries. In summer, "*hygge*" means picnicking in the park, barbecuing with friends, open-air concerts, street parties, and biking — all in the Danish way. No matter when and where you are, candlelight often forms the basis for the *hyggelig* feeling. I'd guess that the Danes consume, on average, 6-8 kilograms of candles per year.

Hygge living in the spring

Flowers and flowers

Spring decorations without flowers? Unthinkable. And because the Danish *hygge* style is based on naturalness, you pick wildflowers, grasses, and herbs for flower decorations and even bouquets from the meadow. If you do not have the opportunity, there are beautiful bouquets from the region on the farmer's market. Flower vases should be plain and simple. They look best on windowsills, wooden tables, or rustic shelves. Exotic flowers, ornate vases, or artificial plants should be avoided or used sparingly. For this, natural woods, for example, beautiful driftwood from the last walk on the lake, are suitable decorative elements.

Blueberry branches for spring decorations

Are you okay with something out of the ordinary? Then place a vase of blueberry twigs on your desk or on a side table. This plant is typical of the north and can also be found in our forests. Sometimes the branches are even available in flower shops. They look best in a simple vase that's color-matched to the rest of the decor. Blueberry branches are an unusual eye-catcher that you can get cheap or even for free.

Home textiles: sun, spring, lightness

Dark blankets and pillows should be washed or cleaned and packed in the closet until next fall. Bright home textiles provide comfort in the spring. If you like cheerful flower patterns, don't overly crowd your space with them. So a colorful pillow could be combined with several monochromatic ones. Instead of heavy, brown blankets, try cuddling up in the spring with light plaids in light gray, beige, or other pastel shades.

Even heavy curtains should now give way to light and airy fabrics. Simple, white or pastel-colored curtains are best for spring. They let enough light into the living room and protect you at the same time from prying eyes.

Hygge Feeling in the Bedroom

"*Hygge*" also means "decelerate," and where

could you do that better than in the bedroom? Provided, however, that the room does not serve as a storage room and is lovingly minimalist. To make sure you can really relax in bed, you get more than just a pillow. It may also be a whole pile. The fluffy flannel bedding is now stowed away in the closet. For spring bedding, look for quality and design. Cotton satin is easy to clean and is comfortable even on warm days. The *hygge* style is best matched with subtle floral patterns and plain or striped duvet covers. White linen is again modern. Patches of color are, of course, allowed. A colorful spring bouquet or individual, tasteful accessories can provide a nice contrast.

Candles and hygge — two things that belong together

Artificial light is appropriate but hardly cozy. That's why the *hygge* style of living relies on candles and otherwise turns on only those light sources that are absolutely necessary. Candlelight brings moments of calm and coziness. Make use of a lot of different candles, preferably in several different sizes next to each other, but say goodbye now to the candlesticks still on the table or bookshelf from winter and Christmas. Simple tea lights and white candlesticks with the modern but

not intrusive design of wood, metal, glass, or stone —
all this harmonizes with your new spring decor.

Hyggelig spring ideas for the kitchen

If you are lucky enough to own a kitchen, you can also decorate and decorate them. In Scandinavia, the emphasis is placed on making a kitchen, not just practical, but a space that offers space and the necessary ambiance in which family and friends like to cook, eat, and sit together. *Hygge*-style kitchens are bright, preferably white, with wooden worktops. A big table is a must. Of course, plants are also part of the kitchen. A beautiful spring bouquet or green plants provide a welcoming atmosphere. On a shelf or on a window sill, you can grow herbs in pots and boxes in spring. Accessories make the kitchen comfortable, choose it wisely so that the room does not seem restless or cluttered.

Living Trends for Spring 2019 in a Nutshell

The current colors can be summed up as "ice cream tones," from vanilla to strawberry to pistachio; everything fits. And best of all, they'll stay in fashion all through the year, even beyond spring.

Slow living

It's not new, but it's still there, and it's a great complement to the *hygge* style: the home becomes a personal haven of peace — friendly, bright rooms without junk and unnecessary ballast few, high-quality favorite place. This creates a pleasant atmosphere for relaxing and refueling as a contrast to the often-hectic world out there.

This is how *hygge* goes in the summer: Here are your best ideas for more *hygge*-feeling.

Campfire romance

Phew, how trite does that sound? Yes, that may be. But this triteness is nice and says something about *hyggelig*. You can do it with guitar music and marshmallows (little treats).

Barbecue with friends

Well, if you love the campfire romance, you certainly would not say no to a cozy BBQ.

Surround yourself more often with your favorite people. Imagine beautiful joint ventures and create such beautiful memories that endure through the summer.

Decorate your garden or patio with lanterns,

candles, lanterns, and garlands. Sure, not all at once, but do so in such a way that you create a pleasant atmosphere with subdued light.

Wake up for the day and go to a beautiful place where you can enjoy the sunrise. What is essential? A cuddly blanket to snuggle in, a warm drink to warm inside, your favorite human (alternatively, your favorite cuddly animal), and comfortable clothes.

"Me"-Time

Yes, that too is *hygge*. Time for you, time for conscious self-time. Grab a deck chair or hammock, your favorite music, your favorite drink, and then just enjoy doing nothing.

Shut the world out — even more, me-time. Spend several hours or maybe even a whole day without any distraction from your social media channels. Enjoy the time with yourself.

Summer is the time for reading. What could be better than lying on sun-drenched grass and reading a great book? Here are some inspirations for wonderful, heart-warming moments: pick berries, eat berries, and make jam. And you know how delicious it is to eat homemade jam on cold winter days and to think about summer pleasures.

Homemade

Everything homemade has a big *hygge* bonus. Of course, this includes homemade lemonade, homemade cakes thickly covered with fresh berries and fruits, freshly baked bread — and whatever else you think of.

Favorite place

Set up a favorite place on a balcony or terrace. These may include colorful lanterns, flowers, and shrubs in beautiful wooden tubs, a hammock, a lounge chair...

Shooting star counting

There are these wonderfully velvety summer nights that have no end and in which the shooting stars fly just like that. If you are lucky enough to catch such a night, celebrate it until the new day. And if there are no falling stars, star-gazing is also very cozy.

Swimming at night

It's kind of magical — especially if you catch a night that's starry and the moon is thick and big in the sky. For the *hygge* feeling, of course, you need cozy blankets again that you can lie down on after work.

15 Tips & Ideas for *Hyggelig* Autumn

1. Warming drinks are perfect for a cozy autumn evening. It does not matter if you drink hot chocolate, tea, golden milk, or something else — the main thing is that it's warming you on the inside.

2. Candles are an absolute must for a successful *hyggelig* evening. In all sizes, colors, shapes. Wherever there is still a tiny spot. And of course, that you can see her and enjoy her light flickering.

3. Warm light and pleasant light sources are so essential. A glaring neon light is an absolute no-go for cozy *hygge* feeling. On the other hand, hot, soft light sources, indirect light, and dimmable lights are absolutely in demand. Of course, as a supplement, fairy lights are also suitable. Basically, everything that brings the atmosphere is allowed and desired. Many smaller lamps, for example, bring more comfort than a large lamp on the ceiling.

4. Cuddling on a bearskin in front of the fireplace — a *hygge* classic! Happy is who can now fire his fireplace and enjoy the

flickering flames. Just dream away.

5. My favorite place is waiting for me. Set up your favorite place to return to again and again so you can feel good and happy.

6. Natural colors convey a feel-good atmosphere. Try featuring natural colors such as beige, cream, and brown. But ultimately, it's about YOUR comfort at home. And if you love strong, bright colors, then that works too. Maybe not in the entire apartment, but rather as an accent color.

7. Bring a little nature into your home. Chestnuts, colorful leaves, autumn flowers, pumpkins, firewood logs, and moss are great for bringing a little piece of nature into your four walls. So it will be a nice autumnal addition in your home too.

8. Cook together and eat with friends. Food tastes best when prepared together. And how wonderfully cozy it is to work with friends in the kitchen, drink a glass of wine (or two), and then enjoy the meal with inspiring conversations.

9. Game night with the whole family and with friends. Immediately after cooking

and enjoying (or as a separate event) a game night in a cozy round is also very *hyggelig*.

10. Nest-building is announced. And it takes as many pillows and blankets and poufs as possible. There is not too much here. Make it as cuddly and as comfortable as possible. Finally, long and cold evenings are approaching.

11. Time for your favorite book is always. But on dark, rainy evenings, it has its own special charm.

12. A cuddly autumn outfit makes a lot. These are clearly chubby knit sweaters, like in XXL, into which you can really swaddle, thick scarves, which are so long that you can warm even your loved one with it — and of course colorful self-knitted stockings.

13. Enjoy homemade dishes. Not only the food of the treats (of course), but also the baking, the whole preparation and the beautifully covered coffee table belong to this enjoyment. The food of the home-baked is then virtually the icing on the cake.

14. Atmospheric feel-good fragrances. Of course, our nose also wants to be *Hyggelig*. All the smells that cause a pleasant feeling in you, are suitable for it. Maybe you have set up a fragrance lamp *, aroma diffuser *, incense sticks, a room perfume or a scented candle.

15. Being in the here and now. And that includes offline that is the new luxury that we should treat ourselves to much more often. Just turn off the phone for an hour, an evening or even a whole weekend. And experience the world only offline. With all your senses.

13 Cuddle Moments & Reasons to Embrace the Little, Everyday Pleasures

1. Create a cozy corner — to dream, to read, to enjoy, to be simplistic. What belongs in it? Everything that is good for you and that makes you feel good. This can be your favorite chair with blankets or the space in front of the Swedish stove.

2. Grab your favorite book and your favorite drink (preferably a warm drink) and seal yourself away from the world around you. This is super easy — turn off Wi-Fi or switch to airplane/flight mode.

3. Make it really cozy. With thick pillows, cuddly blankets, candlelight, a hot water bottle, an XXL scarf and a sweater, wool socks, cozy music, and aroma lamps/diffusers.

4. Have quality time with your loved one. Just time for you both to comfortably drink a glass of wine, to cuddle, to chat.

5. Enjoy cozy time with friends and family. Cooking together & (even better) eating together, playing a great parlor game again (we played man-annoying-you recently, and it was sooooo funny...)

6. Spend time with yourself and do the things you like to do. What have you wanted to do for a long time? Wouldn't it be a good idea to write in a gratitude journal again, to meditate, to start journaling or scrapbooking, to listen to an inspirational podcast, to dance to your favorite music?

7. Clear your head. Get out in the fresh air whenever possible. Maybe you're building a

snowwoman or a whole snowperson family. Maybe you'll take a walk and discover little wonders in nature. Maybe you'll just keep your face under the sun's glow and heat. After that, going back to the cuddly home is always twice as nice.

8.   Bake a cake and invite friends over to have coffee.

9.   Spend a few hours in your home spa and pamper yourself — with your favorite bath oil in the bathtub, with a fragrant body lotion or a nourishing face mask.

10.   Decorate your home. Bring winter with all its warmth (the cold stays at the door) and comfort in your home. It can come from candles, fragrance lamps, fairy lights, cuddly corners with big cushions and warm blankets, warm colors, soft lighting, winter flowers, and so on.

11.   Sort photos, indulge in beautiful memories, and make a photo collage of the most beautiful experiences, then hang it up or put the photos into an album, so you can experience those beautiful moments again and again.

12.   Consciously enjoy having nothing to do

sometimes. This is so difficult for most, and at the same time, it is so important. Maybe build castles in the air or drill holes in the air.

13. Embrace more — your sweetheart, yourself, your inner child, your sunny attitude toward life, and the little joys of everyday life. Then you have much less time to worry about your everyday frustrations.

4. HYGGE — THE DANISH PHILOSOPHY OF HOW TO BE HAPPY

Happiness represents one of the existential conditions most yearned for by every human being, at any latitude of the planet. Many seek it by accumulating material objects, pursuing fame, money, or increasingly perfecting the external aspect, but happiness, despite numerous efforts, continues to be unattainable.

To experience it, to truly experience it, it is fundamental to cultivate loving kindness, compassion, empathy, controlling disturbing mental states such as anger, malevolence, and cravings. According to the recent World Happiness Report, the happiest people in the world live in Scandinavia. Finland, Norway, and Denmark are the countries where the quality of life is highest. In particular, on Danish soil, happiness is not ephemeral, because it is developed by adopting an ancient and at the same time, modern philosophy, called *"hygge"* (pronounced "hoogah").

"Hygge" can be defined as an art of living that

36

places tenderness, conviviality, and well-being in its entirety at the center of existence. In the Danish language, originally, "*hygge*" meant "to console" or "to encourage."

According to this philosophy, to nurture a state of happiness, it is essential to live by privileging the dimension of Being rather than that of Being (to paraphrase a well-known theory by Erich Fromm), in everyday gestures, and in simple or great actions. This means going beyond the individualistic perspective to embrace, when possible, moments of sharing with our loved ones, friends, and other people around us.

THE SEARCH FOR A SENSE OF BELONGING

Nicolai Frederik Sever in Grundtvig (1783–1872) poet, historian, pedagogue, Danish reformer, laid the foundations for the development of the *hygge* philosophy.

Everything started from his project called "*folkeoplysning*" ("popular education"), which brought together the typical values of the Enlightenment and principles related to personal liberties.

He was also a staunch anti-elitist. *Grundtvig*, also inspired by Nordic mythology, believed that Denmark should seek its own prosperity in the welfare of its people in order to forge a national identity based on a feeling of belonging.

The vision of *Grundtvig*, therefore, focused on the internal development of the nation and not on the search for external greatness to the detriment of other peoples (as did many other nations, not only Europeans, between the nineteenth and twentieth centuries).

It was thanks to this man so visionary that the popular schools were founded, which also allowed many peasants (who constituted the majority of the Danish population until the beginning of the twentieth century) to follow literacy courses.

In addition, national and popular education was stimulated through choral songs, poems, and stories.

Oral traditions glorified ideas such as simplicity, joy, community, solidarity: all elements that distinguish them and their culture.

MOMENTS OF SHARING

Living according to the *hygge* philosophy means cultivating and expressing sweetness, searching for situations in which it is possible to feel truly connected with others, as Louisa Thomsen Brits, author of the book *Hygge*, states that "the Danish method of daily pleasures" is we all need to reconnect with authentic and solid relationships, especially in an era as unstable as the one we are experiencing. It is, therefore, vital to go beyond the virtual, beyond the internet, to find concrete human relationships.

SMALL, SIMPLE GESTURES

Embracing this philosophy is not so difficult, even beyond the borders of Denmark. It is not necessary to perform sensational actions or deprive yourself of anything. What really matters is remaining open to others.

When we overcome a closed and limiting way of life, we learn to marvel at life and to savor it fully. In this sense, even apparently simple moments turn out to be acts that are no longer ordinary, but extraordinary, such as preparing a cake with one's daughter/child, or organizing a picnic in nature with friends, or dining by candlelight with your beloved.

These situations, if lived in full awareness and with an attitude of openness, become unforgettable moments.

At any time, we can see the beauty and strength of life.

THE PLEASURE OF THE PRESENT MOMENT

Living according to the *hygge* philosophy means savoring the here and now.

This is a common principle of millenary Oriental wisdom. In fact, if one continues to brood about the past or if one is afraid of the future, one does not live life to the fullest. Indeed, it is impossible to grasp the poetry and intensity of the present moment.

In this regard, it is interesting to note similarities between the *hygge* philosophy and meditation practice. In fact, one of *hygge*'s principles is to "slow down actions," performing them in a conscious and gentle way, without haste, without nervousness. With the right conscious attention, everyday behaviors are carried out, whether they are important or not. For example, every time we walk, we can listen to our inner sensations, we can perceive external stimuli, and we can understand how our body reacts to them.

LIVING IN AUTHENTICITY

In a hyper-connected world, the virtual often prevails over reality, and one runs the risk of losing one's points of reference, one's essence, one's identity.

To enjoy true happiness, it is essential to remain anchored to reality.

It is important to reconnect with one's inner world, living in harmony with one's values. The *hygge* philosophy suggests that you stop every day to observe your existence, to hear and collect, even in writing, the thoughts and feelings that occurred during the day. Listening to the voice coming from our inner self allows us to remain anchored to reality, to our values, and to our authentic needs.

CONNECTING WITH NATURE

To cultivate authenticity, bucolic places come to our aid, where they can relax and stop. Those who continue to adopt frenetic rhythms risk losing themselves in the many inputs from outside, from ephemeral fashion, from propaganda slogans, not only from social conditioning. It is now established scientifically (see the studies of Dr. Qing Li, immunologist, among the greatest experts in the world of forest medicine), that walking and immersing yourself in nature — in particular, where there are trees, then in a forest, in a park, in a pine forest, etc. — is beneficial for our body, since the biochemical substances (terpenes) released by plants stimulate our immune system.

In fact, when we breathe in terpenes, the amount of natural killer cells in the body (those that help eliminate viruses, bacteria, and other pathogens) increases dramatically.

Hence the importance of disconnecting from social media and the web to connect rather with nature. The Danes, when following the *hygge* philosophy, get their fill of light and vitality by cycling or walking in the beautiful forests of Denmark.

Everyone can re-establish their relationship with Mother Nature; in doing so, it helps our body to feel better, allowing our mind to relax.

We also do good to nature itself by connecting with it, because we cultivate love and respect for the earth, all plants, and every creature. In this way, we feel part of a single, great, living organism, which is our planet.

CREATING THE RIGHT ATMOSPHERE

When we talk about "*hygge*," we must take into consideration what the Danes call "*stemning*," or the right atmosphere. Simplicity, softness, and harmony are the elements that distinguish a *hyggelig* environment. You can create a *hyggelig* corner in your home, where you can relax after a long, tiring, and stressful day. For example, if you have a small sofa available, you can easily move it next to a window, so that when you lie down, you can admire the sky. A *hygge-focused* environment prefers candles, rather than artificial lights, furniture, and natural fabrics, soft colors, to facilitate restfulness and calm. At your "corner of peace," you can add incense or diffusers for essential oils, choosing fragrances that make us feel good.

5. THE DANISH MENTALITY

The Danish mentality is something subtle; it is both highly appreciated and repulsed at the same time. It is something very special — and we know it well!

Danes are informal people who put equality next to coziness.

Discussion and argumentation are fundamental elements of Danes' upbringing — in society as well as family. This means that we Danes have a fundamental awareness that it is possible to be heard and make an impact. In the association life, you can become acquainted with the democratic processes and structures that characterize Denmark. The associations have social, informational, and democratic influence. This democratic process, which we get through the breast milk, is something very special and Danish. It is this everyday democracy that greatly contributes to shaping the Danes' mentality and awareness of themselves and their fellow citizens.

Hygge is a very important element of the Danish mentality. It may be difficult to translate the concept of coziness, but one will quickly discover that coziness is closely related to being comfortable with one another and with food and drink. *Hygge* is a Danish brand and something that we are very proud of. It is said that if the Danes exported coziness like bacon and beer, then the world would be a much nicer place.

This coziness is very difficult to explain to outsiders. Many Danes have tried in vain to explain that it is about creating a special harmonious mood, which can be produced by various aids such as candles, good food, and heat. Having fun is a concept we Danes can be proud of. I myself almost explained the term to a bunch of Scots who now understand the essence of the term and diligently use it. I'm a little proud of that.

Humor is also an essential element, and for many Danes, humor includes a great deal of irony. At first, it may be difficult to understand the irony, but it is an important part if you want to get into the mentality of the Danes. Even many Danes believe that our humor is absolutely exceptional and wonderfully good, while others think it is flat, and often at the expense of others. In particular, it is our immigrants who often have to face our plates. Many defend this criticism by saying that our guests simply have to get to know the Danish culture because now they are here at our expense. Shouldn't we Danes be magnanimous enough to get a little bit into our fellow man's situation, and thus learn to hold back a bit with our jokes?

Welfare secures citizens to a considerable degree financially, e.g., in connection with illness, unemployment, and old age, in addition to supplementary measures, including support for housing and children expenses. There are also a number of highly expanded services in the form of daycare centers, health care, home care, etc. I think the Danish welfare model is something that we can be very proud of. It reflects the idea of charity, and of wanting the best for everyone in society — including those who cannot help themselves.

The principle behind the Danish welfare society, which is often referred to as the "Scandinavian welfare model," is that all citizens deserve access to social benefits regardless of their social background or origin.

Foreign observers often find that Denmark, or at least Copenhagen, is a safe place to move. I think many foreigners appreciate the sense of security they experience in Denmark. Foreigners who come to Denmark often emphasize safety and security as the most important characteristics of the country, e.g., children go to school alone, and even well-known business leaders need not surround themselves with security guards. The parliament is open to everyone, and it is not uncommon to meet ministers on a bicycle on their way through Copenhagen. Even the Queen can handle her shopping in Copenhagen with only minimal security service.

This image of security is not just fiction — it is reflected in the statistics, which shows that the crime rate in Denmark is among the lowest in the world.

Mentality and culture go very close hand in hand; it is these concepts that define how we are as human beings. It would then be a paradox to think all Danes think and react in the same way. We must see that there are disparities and differences in society, and be proud of that diversity. There are white, yellow, black, brown, and probably some totally fifth colored people in Danish society. We are all part of creating the Danish culture, and this is a reality that we must learn to accept. Culture is something that is constantly influenced and changed — the Danish mentality was very different 100 years ago. That is why we should not be so afraid of all the new happenings in society. Instead, we should try to understand it, and then live in the present!

6. HYGGE FOR YOUR HEALTH

Benefits of *Hygge*

There are different benefits associated with *Hygge* practice. The researchers of happiness are finding out again and again that Denmark has some of the happiest people on Earth who the Danes attribute to *hygge* practice. Increased happiness might be an advantage to practicing *hygge*, but there may be other emotional, physical, and relationship benefits.

Emotional benefits

Hygge decor is designed to promote peace and quiet in the living space. Understanding our experience and environment through the use of seeing, hearing, touching, tasting, and smell, it's no surprise that creating a cozy living space makes us less anxious and encourages a sense of emotional well-being. His and security. This sense of security and security can better allow us and those who share the space with us to lower our guards and be more up-to-date and open to connecting with each other.

Examples of possible emotional benefits can

be:

- Less depression and anxiety
- Increased feelings of self-esteem
- Increased optimism
- Reduced stress
- More mindfulness
- Improved self-compassion
- Heightened gratitude

Physical benefits

In moments of perceived danger or threat, our bodies naturally go into a reaction of struggle, flight, or freezing. A *hyggelig* environment promotes an atmosphere of safety and comfort where the body and mind can feel more relaxed. In such an environment, there is less need to search our environment for threats.

Examples of possible physical benefits can be:

- Sleep improved
- weight regulation
- Less cortisol (stress hormones) peaks
- The improved practice of self-care
- Reduced need for unhealthy coping behaviors such as alcohol or recreational drugs

Social benefits

If we feel comfortable and emotionally secure, we will seek to build and nurture relationships with others. In a *hygge*orientierten lifestyle, value is attached to the connection with family, friends, and relatives. We feel more confident when we are connected with other people, we feel safe taking risks, and we are more open to practicing vulnerability with others, all that is possible in a *hygge*-like habitat.

Examples of possible social benefits can be:

- Concentrate on being together
- The feeling of comfort and safety
- Increased confidence
- Increased intimacy
- New social connections
- Existing relationships improved
- Less dependence on social media

So use hygge in your life!

Most of us would like to feel happy, peaceful, and comfortable, but must we move to Denmark to take full advantage of the *hyggelig* lifestyle? No! There are many ways in which we can integrate elements of *hygge* into our daily lives and living spaces. Implementing some of these elements can begin to give you a feeling of peace, connection, and comfort in your everyday life.

Lighting

Lighting is an integral part of creating *hygge*. The use of warm, soft white light creates a welcoming and comfortable space compared to hard, bright white incandescent or fluorescent lamps. Remember, the higher the lumens on the bulb, the brighter the light. You can also install a dimmer switch to have options for lighting the room to your liking.

Another method using floor lamps instead of ceiling lighting. Ceiling lights can produce light that is too bright for the room and feels institutional. By using floor and table lamps, the lighting can create a more intimate space and areas where people sit and read, relax, and talk.

After all, candles are a striking light that is used in a huge space. Candles naturally create a warm, soft light and a sense of relaxation and comfort that is particularly popular in this style. If open-flame candles pose too much danger to your living space because of pets or children, you can use LED candles instead.

Texture

Hygge living prefers things that feel soft and comfortable. You should use soft accessories such as blankets, throws, pillows, and rugs to create a warm, inviting space. The soft textures soothe and calm us as our fears become high. Soft textures allow others to calm and feel safe in the room, to calm fears, and to enable people to be more open to each other. Conversations in this room can feel calmer and more open than feeling hectic or under pressure.

Decor

A soothing environment can be created with accessories such as wood trim, houseplants, and simple, clean decor. Use pieces that have a special meaning, e.g., pictures of family members and relatives. You can put photo albums with pictures of travel or experiences shared with others on the coffee table.

Warmth

Heat is not so much about temperature; it's about a feeling of emotional warmth. A fireplace is a hallmark of a *hygge*-style room, but that's not an option for everyone. Everything you can do to create this inviting, warm space will be a plus. Examples include candles and displays with accent lighting. You can also use small fairy lights in certain areas of the house to create the inviting warmth that a fireplace gives the room.

Color

The colors that are chosen for a living space are an essential part of creating a cozy stage for you and your guests. Often, neutral colors are chosen, especially white, soft white, rouge, and soft brown. The use of minimal, neutral colors on the walls actually helps to calm your mind, which fits this particular style of Living. *Hygge* is calm, soft, soothing, and comfortable.

People

The goal is to be present and to connect with the people around you. By cultivating these relationships, we allow ourselves and others to experience a sense of belonging. If we are, then we feel emotionally safe. Emotional security, on the other hand, creates a positive social experience and allows us to feel the physical benefits of lightness, calm, and connection.

Activity

Hygge activities usually include things that help us

feel peaceful, comfortable, and connected to others. Meetings with friends at home are a major activity. The meetings focus on connecting with others, not the presentation. There is no need for a formal affair. *Hygge* living would indeed suspect the opposite. Meetings should provide a space that is casual and inviting, providing people with a place to feel good and to focus on relationships and connect. Consider a game night with friends, friends, or neighbors for coffee or a book night.

Hygge at Work

Work is a place where it can be difficult to take care of ourselves or create an environment that is focused on peace and tranquility. Find small, safe ways to create changes in your space to promote a sense of calm. You'll find it easier to enjoy the space, which could lead to better productivity and job satisfaction.

Ideas to try:

- Accent lamp with soft white illumination
- Small, potted succulent
- Reasonably sized rug for your work area
- Pictures of family and friends

A Word from Verywell

Hygge is about comfort, peace, and connection. The benefits of some of these elements affect our emotional health, physical health, and social health. By incorporating some of these ideas into your living environment, you can provide a relaxing, inviting space for your health and well-being.

7. WHAT IS BEHIND THE *HYGGE* ENTHUSIASM?

For several years, *"Hygge"* and *"hyggelig"* have been on everyone's lips. But what does that mean? It describes a way of life that wants to enjoy life with simple means to the fullest. The emphasis is on "simple."

The Danes have realized that to be happy does not require much. On the contrary, often, it helps to get the right thoughts.

If these do not come naturally, they can be consciously guided in the right direction with a little bit of routine. For example, by deliberately directing one's gaze from the negative and the stressful to those aspects that work well in life. And you're smiling again. Ultimately, it's about making life as comfortable as possible for yourself and others.

There are ways and means for this, which are reflected in different areas: health, housing, living conditions, and authenticity.

Health

Eat healthy with quality foods to feel good in

your body.

Housing situation

To make your own home as cozy and comfortable as possible, create a feel-good climate where you can relax and unwind —candles/scents, pillows, seating, beautiful carpets, and pictures help.

Living conditions

A well-balanced work-life balance: often, an after-work beer with friends, sports, or a visit to the cinema works wonders after a stressful working day and provides a necessary balance. What the Danes certainly do not do: sit on the couch and ponder.

Authenticity

To stand by yourself and only do things that you really stand for can relieve you immensely.

Advantages of the *Hygge* Style

The *hygge* style has many advantages: Denmark, Norway, and Sweden are among the happiest people. With only a few tools, a person can create and do great things: those who feel comfortable in their environment are equally more balanced and productive. Then it is not that hard if it takes longer.

Being deliberately anchored in everyday life, plus mindfulness and self-esteem, contributes to higher well-being and increased fitness, making both more stress-resistant.

Mindfulness and self-esteem in everyday life

Granted, that is anything but new. Nevertheless, the *hygge* movement has triggered a huge hype. The topic is moving! *Hygge* is the trend, and the corresponding guides and products are catching up.

This is because the ability to be mindful and self-respect has largely disappeared, especially during the study and professional life. Many are obviously asking, "How can I go back?"

The current facts speak in any case for this theory. Especially in the modern working environment, speed and pressure are constantly increasing as a result of digitization and the lack of skilled workers.

Even student life has not gotten simpler: many surveys show that students today are under more pressure than before. They themselves are responsible for surveys for stress in their studies or job and increasing pressure to perform psychological problems.

Is your business *Hyggelig*?

Hygge cannot cure depression

To make that clear from the outset: *hygge* does not help cure major depression. For this, sufferers must necessarily go into the hands of experts. But *hygge* can cushion the melancholy or help it not to let it develop in the intensity by the right way of life and attitude.

Workers and students should learn to break specific behaviors that do them no good.

For example:

- Consciously and purposefully lead to good conversations.
- Cancel calls that are not effective.
- Looking forward to small successes.
- Consciously perceive positive situations.
- Do not focus on negative situations more than the positive ones.
- Pay attention to the signals of the body.
- Regularly enjoy the environment and nature — fresh air often brings clarity and calm to the head.
- Surround yourself with people who mean well with one.

Tips: How to Get Your Business *Hyggelig*

What should have become clear: *hygge* is not fixed teaching, but more a way of thinking that causes much positive with consistent application. But where do you start? *Hyggelig* feelings can be anchored practically in everyday life. For this, you should not immediately fall into wild action.

Some men do not have a lot to do with the "deco bustle" and kitsch, but rather, they value minimalism. That's okay, because...

a) everyone has their own taste, and

b) *hygge* can come in many forms and senses.

From the following tips, you should, therefore, pick out what suits you personally:

Hygge for living and working

More and more employees have the opportunity to set up their workplace to a certain extent, according to their own ideas. Students have it even easier in their shack/dormitory room. *Hyggelig* makes it easy with such accessories, which you can use for home living as well as for work in the office, for example:

- pleasant room fragrance
- fresh flowers

- colorful light-change lamp
- delicious teas
- nice cups
- great pictures

The desk and walls can be brightened up with things that arouse positive memories and make daily work a little more cuddly. They act as a kind of *hygge* reminder, reminding you to approach yourself with greater awareness. This requires some practice at the beginning but eventually gets you in the mindset.

Hygge for well-being

For example, you can do something good for your body by leaving *currywurst* in the canteen or cafeteria and eating a delicious salad instead.

For acute stress, this is advisable: consciously do not choose overtime or night shifts during the course of studies, but to make sports the necessary balance in the evening, then go to bed early, and the next day, you will go to sleep and feel well-rested.

Often, things go to sleep after a cap and with a little distance much faster and easier by the hand. Incidentally, the Danes are very consistent in this regard: Here, notorious overtime workers do not reap appreciative looks but, rather, sympathetic looks. Follow this motto: Tomorrow is a new day!

Hygge through fitness

Physical well-being is not only related to sleep and restiveness. *Hygge* followers propagate the closeness to nature. They can be anchored in their professional life by deliberately turning around the block after the lunch break, instead of inhaling smoke in the smoking-room. And that movement works wonders, releasing endorphins.

If you prefer to take the stairs instead of the elevator, you will also get your circulation and psyche going. The same applies to the way to the university or the workplace — rather, leave the car and instead take to foot or bike to reach your destination.

Hygge through planning

Good time management is very important for students and employees. There should always be room for breaks in your own calendar. In no case should a to-do list take over your life.

Setting generous deadlines helps keep you relaxed before you get stressed again. Solid prioritization is important. It may require you to refuse help requests from friends or co-workers.

Do not worry: saying no only hurts at first. In

the long run, it can even gain you respect. The same applies to students: they should start early to work on chores and papers and not just start shortly before the deadline — that takes a lot of pressure off.

8. HYGGE IN THE OFFICE: PRODUCTIVITY MEETS COZINESS.

A cozy atmosphere that contributes to communal exchange and harmony sounds more like a living room than a workplace. That's exactly what will change in the future. Because the current labor market change also leads to renewal and redesign of workspaces, coarse carpeting, open-plan offices with partition walls, and little room to breathe are common problems. Productive work today looks different. Perhaps this is why the Danish *hygge* is so well-equipped for many offices: with minimalism, open forms, lots of wood, and visual lightness, *hygge* now moves to the desk space. How exactly can that look? We took a look at the interiors of modern offices: it gets *hyggelig*!

Bright, homely hygge.
Hygge stands for a way of life and quality at the same time. In Danish culture, this form of coziness is an antidote to the long, dark winters. Because even if it is freezing for several months,

at least your own four walls can radiate heat. It's best to help against the perennial noses with a log fire, lots of candles, warm blankets, a purring cat on your lap, and lots of good friends. *Hygge* describes the ability to do it comfortably with simple means. *Hygge* can also be an evening to be fed and drunk together in a familiar round, and the leisure comes to good conversations. A feeling of security (and a sauna session afterward) sounds like a vacation, right? The *hyggelig* holiday feeling from the private living room has long been extended into public life because a *hyggelig* interior communicates this attitude to life in cafes, restaurants, hotels, or offices.

Hygge in the Office: Five Tips

Danish design stands for clear, light shapes and colors that are unobtrusive and open up space for ideas. In the office, this *hyggelig* lightness can positively affect the productivity and the community feeling of the coworkers. With the following inspirations, *hygge* can move into every office:

Brightness!

Natural light is important for concentration. Large windows and plenty of light suggest space and create the right ambiance for new projects. In-office planning, the supply of natural light sources should, therefore, be a central design element.

Meeting places!

Hygge stands for collaborative exchange. Be it summed up the last meeting, lunch break, a brief rest period, or a small talk among colleagues: office space should provide room for the gathering. Seat islands to relax bring. As a change in the daily work, allow new perspectives and yes — leave room for new ideas.

Open, uniform workplaces!

Partitions literally lead to a feeling of boredom. Who wants to feel like a lying battery at work? Communal, long work surfaces structure the space in a concise manner and take nothing away from its vastness. This is important so that your eyes can walk around to recover from the screen. Also, co-working concepts are well suited for the *hygge* vibe, because here, they all stay on the move and now and then have a new face next to them.

Individual needs adjustment!

Community and personal arrangements fit

together very well. Provided the right infrastructure is created for this. This means that chairs should be adjustable for different heights, high tables are available, and the own display can be adjusted in relation to your own body size. Lamps, if also desired with daylight function, should be adjusted individually, and the feet can sometimes be raised. All this is possible in a few steps — and therefore all the more *Hyggeliger*.

Less is more!

Your own workplace should not feel like a motto party. Soft, dull, and few splashes of color harmonize well with a clear division of spatial functions and organic materials such as wood and (real!) plants bring heat into play for a lot of natural coziness — *hygge*, of course.

Workplace or adventure playground?

A modern work organization goes hand in hand with a modern facility. Especially large, international companies use the design potential of their office space to express their corporate Identity. This makes an impression on international customers and looks beautiful at the same time. But well-thought-out workplace concepts should especially benefit their own employees. With modular seating, innovative room layouts, and functional design, it quickly becomes *hyggelig* in the office.

In this context, *"hygge"* stands for the feeling of having room to breathe, to think, and to exchange ideas. In order to create this atmosphere, some companies take deep into the bag of tricks. A treehouse as an office? Hanging chairs and beach bars? Or a clearer view of the room that invites you to linger? From adventure playground to relaxation pool.

9. HOW DO YOU MAKE AN OFFICE COZY?

1. MAKE GOOD USE OF BASIC SUPPLIES

To create a cozy office, you do not necessarily have to purchase all kinds of plants, pillows, and trash cans. It is precisely by making good choices when purchasing facilities you already need that you can make a difference. For example, office lighting is often cold and the same throughout the workplace. By working with different light sources, you can already create a completely different atmosphere in the office. Furniture, such as file cabinets, is also needed anyway, but depending on what kinds you choose, they can make a big difference in the look of your office.

2. SOCIAL PROVISIONS

A good atmosphere is not only in the appearance of your office but also in the options for dealing with colleagues in a pleasant way. For example, is your coffee corner inviting? Are there options for informal meetings in sitting areas? When employees have the opportunity to engage in informal discussions about work and other matters, your office will appear more open and pleasant.

3. CHOOSE A COLOR SCHEME

A workplace generally needs a lot of things. When you order folders, notebooks, pens, etc., functionality is, of course, the highest priority. However, make sure that you also think carefully about the colors of the items you purchase. With a warm color scheme, for example, your office radiates comfort, which immediately benefits the atmosphere.

4. PERSONAL APPROACH

Something to think about is the extent to which employees themselves can create an atmosphere in the office. Photo frames and your own plants make a workplace more homely but can also make desks look messy and cluttered. Fortunately, it is not an all-or-nothing choice, and it may well be possible to draw up rules about the design of desks and still give employees the feeling that they have something to contribute.

5. PETS AT THE OFFICE

This will not apply to every office, but if you really want to give your office the feel of a house, a pet can be a good choice. A dog in the office provides a connection among employees and makes the prospect of going to work in the morning a lot better. However, this is only suitable for companies that can really offer the pet a nice life and where none of the employees object to it being there.

10. HOW TO CREATE A COZY WORKPLACE

When you think of a standard office, you probably see a soberly decorated room with bare walls, cold colors, sharp corners, cool white light, and artificial materials. If you stay in such a room for too long, it can affect your state of mind and well-being and therefore your performance.

It is, therefore, not surprising that a major change is taking place in the field of office design, and the trend towards a more intimate workplace is growing.

When designing new workplaces, interior architects not only pay attention to functionality, but also to the comfort that it offers to the users.

WARM COLORS

Adjusting the color palette is a simple solution that can have great results. To create a cozier atmosphere, it immediately makes a difference if you choose warm colors instead of chilly ones.

You do not immediately have to take drastic measures and have the entire office repainted, but start with one wall or part of it. Warm colors are not only recommended for the walls, but also for furniture and accessories. If you want to make a room cozier without major renovation: put down beautiful flowers and add colorful accessories. Simple, fast, but very effective.

LIGHTING

Lighting determines the atmosphere in a room and largely determines how a room comes across to you. Cool, white light influences your state of mind: it stimulates and promotes concentration. It is therefore widely used in offices and in other places where people need to concentrate on their work. Warm colors, on the other hand, have a relaxing effect and immediately give a room a cozier appearance. In offices, this type of light is used more often in areas such as the company restaurant/cafeteria and relaxation areas, like the break room.

Well-designed lighting with the right intensity and hue can make your workplace cozier and more comfortable and makes you feel more comfortable. Did you know that good lighting and sufficient daylight are linked to a 15% reduction in absenteeism among employees, and an increase of 3 to 20 percent in productivity?

FUNCTIONAL ACCESSORIES

If you want a cozier place, it is a good idea to add accessories to the interior. Colorful paintings on the wall, soft pillows, blankets, or ottomans. Rugs and carpets in different colors, shapes, and patterns also give your interior a unique character.

Plants on the desk or in beautiful flower pots bring life to the office. The role of accessories does not have to be limited to "just looking nice;" they can also be functional.

For example, a beautiful lamp is not only a source of light, but it also influences your feeling of well-being, and colorful soft poufs invite you to a much-needed relaxation moment during a busy day.

With only a few minor adjustments, you can give your workplace a friendlier appearance, and this makes it more pleasant to work in. Research has shown that a comfortable workplace for 91% of employees determines how well they feel in the workplace.

Natural materials, interesting patterns, practical accessories, warm colors, and good light will make your office feel like home. A room that has been decorated in this way will be a nice place to work and meet colleagues and customers.

GOOD OFFICE LIGHTING AS A WORK VITAMIN

Although employees are statistically more likely to complain about noise in the office, this does not appear to be the main cause of reduced work efficiency. Office lighting has a much greater impact on the productivity of employees.

Try to answer this question: would you like to work in an office without windows?

VITAMIN D INCREASES PLEASURE AND PRODUCTIVITY

People who mainly sit behind the computer during their working day often complain about the daylight that reflects on the screen.

Because of this, even when the sun shines, IT departments generally close the curtains or blinds and work in a semi-dark office environment.

Can you imagine working in these circumstances on a daily basis, without the possibility of looking out the window and being devoid of natural light? In the summer, this is not even such a big problem.

At the end of the day, you exchange the artificial office lighting for a portion of vitamin D, and you still have plenty of time to enjoy the fine weather. But what about the winter months?

OFFICE LIGHTING: PREVENT COMPLAINTS AND STAY ALERT

Sunlight affects our biological clock. A lack of sunlight can cause a disruption of the daily rhythm and, ultimately, insomnia or permanent fatigue. And these complaints, in turn, have a significant impact on employee productivity.

Because artificial lighting disrupts the natural rhythm of people, it is of the utmost importance to place the office lighting component high on the agenda. This way, everyone benefits from a pleasant, comfortable working environment in a fresh, fit, and focused way.

THE SPECIALIST KNOWS THE POWER OF LIGHT IN THE OFFICE

When designing an office, it is worthwhile to rely on and build on the knowledge of specialists such as architects and interior designers. For them, natural lighting and artificial office lighting play an absolute leading role in the design of an office or interior. This valuable knowledge helps you to avoid situations where the light in the office is insufficient or causes complaints.

SHARING A WORKPLACE MAKES THE ENTIRE OFFICE WORK

Our research's results show that the average occupancy rate of the workplace is around 45 percent. Are modern organizations able to finance and justify this amount of unused space?

Companies are increasingly considering implementing workplace-sharing within the office environment. With this, they not only take into account the financial side of the story, but they also seize the opportunity to make better use of free space.

You arrive at work in the morning, enter the office and walk to your locker to put away your personal items. With your laptop and a few documents at hand, you start your working day, which, as always, starts with a consultation with your manager.

You go directly to the meeting room that you reserved. After the meeting, you have a phone appointment with a customer. On the spacious sofa, you first go through your notes to prepare yourself for the interview, and then you discuss the latest state of affairs with the customer.

After the conversation, you want to answer a few e-mails. You must also prepare a report. You do not need complete silence for this, so you go to the general office space. You take a seat at a free desk, connect your laptop, and open your inbox.

PROFIT, NO LOSS...

That is how your workday may look in part in an environment where the workplace is shared, a solution that often encountered in flexible offices. Nowadays — a time when the trend of sharing different things such as cars, accommodations, and services is popular — it is no longer a crazy idea for different people to share a desk.

Yet this raises a lot of fear and doubt. This is because sharing a workplace is often incorrectly associated with losing one's own workspace.

In fact, sharing the workplace is a broader vision that offers various facilities and benefits the employee.

According to this concept, the efficient and comfortable execution of office work is not dependent on having your own desk, but rather on the possibility of having access to different places where specific tasks can be performed.

Therefore, when implementing this solution, it is necessary to develop additional work zones, such as places for making phone calls, spaces where you can work undisturbed, informal meeting zones, meeting rooms, and creative or project zones.

"HOT BUNKING" AS A SOURCE OF INSPIRATION

Where does the idea of sharing a workplace come from? The inspiration behind this concept has its origins in the maritime world, where "hot bunking" is a well-known principle.

This term is often used onboard Navy ships and submarines. The space on these ships and vessels is expensive and scarce and is mainly used for storing weapons.

Because sailors work in shifts and part of the crew is always on guard, there are fewer sleeping places (cages) than people on the ship. The crew members alternate using the same bed, which means that the beds are always occupied and warm — hence the term "hot bunking."

A BETTER INTERPRETATION OF AVAILABLE SPACE

The number of available sleeping places depends on the shift work system on the ship. In most cases, there is one bunk bed for two or three crew members.

"Hot bunking" makes it possible to minimize the space required for sleeping, while the saved square meters can be used for other purposes. This idea of sharing available space was adopted by the office world under the name of "hot-desking" and came in several variants in the 1980s and 1990s. However, implementation was hampered by the limited availability of technologies and a lack of tools that allowed employees to work effectively when they were not at their assigned or fixed workplace.

HOT-DESKING AND WORKPLACE SHARING: THE DIFFERENCES

Nowadays, modern technologies offer great flexibility, and there are many forms of room sharing. Two of these forms, hot-desking and workplace sharing, are the most popular within offices. It is, therefore, worthwhile to discover the differences between these solutions.

HOT-DESKING

One agency that is shared by people who perform their duties at different times. This form has been specially developed for "mobile" employees and employees who work in shifts, for example, telephone advisers.

An office often has a specific area for hot-desking assigned to employees from other departments — another name for this form of sharing workplaces is "hoteling" — when colleagues want to work together on a single desk for a project but cannot do that at their regular desk, because the classification does not allow this.

SHARING A WORKPLACE

Temporary occupation of an available workplace by an employee is a form of utilizing workplaces intended for departments in which meetings are regularly held, both inside and outside the office. This often results in a large number of unoccupied workplaces and desks.

Within the vision of sharing the workplace, employees do not have a fixed, allocated place, but use the entire office and the different zones during their workday to perform work.

If a certain employee has to or wants to work at a desk, then they can use one of the free workplaces. When the work is done, it is a matter of cleaning up and leaving the desk tidy for the next colleague.

Depending on the needs or project requirements, employees can also rotate and use the workplaces in different zones of the office. The number of workplaces for a specific department or team depends on the office space's size, mobility, and flexibility. It must be measured before you start shuffling people around.

THE STRUCTURE OF THE CONCEPT

A desk is not always the best place for different tasks at work. That is the most important underlying idea of sharing a workplace. It is not normal to sit at a desk for eight hours and perform all tasks equally efficiently.

Let's look at how we work at home. When we are working on an important project, we usually take a seat at a table. We read a book on the couch or in an armchair. If we want to come up with new ideas, we assume a comfortable position.

Contemporary concepts also promote this way of working in the office. In the past, we did everything at a desk: making phone calls, writing e-mails, going through important documents, and developing new ideas. Nowadays, places and spaces are designed for the optimum conditions for performing all these specific tasks. The agency no longer plays the multifunctional role that it did for years.

EMPTY DESKS = UNNECESSARY COSTS

Economic aspects form the second principle for workplace-sharing. Empty desks are the hallmark of contemporary companies that allow flexible forms of work but have not adapted their offices to this. According to our analysis, the average occupancy rate of workplaces within companies is 45 percent. This means that, on working days, more than 55 percent of the workplace remains unoccupied. Every unused square meter generates unnecessary costs for the employer.

Sharing workplaces and even sharing desks can reduce this problem. There are fewer desks than the number of employees, which saves space—space that, for example, can be used to divide the office into different zones.

IMPLEMENT DESK-SHARING WITHIN YOUR ORGANIZATION

The introduction and implementation of desk-sharing is a huge challenge for a company. It is about changing habits, the work philosophy, the management style (staff is being distributed more), the flow of information, and even communication methods. Sharing workplaces also requires technological changes and redeployment of office space.

SHARED WORKPLACES: MEASURING IS KNOWING

The decision of whether or not to implement shared workplaces within a company must be based on a series of analyses.

The way of working for individual units within the organization and the occupancy rates of agencies within certain departments must undoubtedly be investigated.

Detailed data is also important for the next phase: designing the office space.

This is a crucial project, because the office layout must be tailored as closely as possible to the character of the company. Employees, the target group, and the people who use the office every day must have access to all the necessary work zones in the new office space.

Moreover, the number of zones, all optimally located, must be correct.

PREPARE EMPLOYEES FOR THE NEW STYLE

Complex and difficult processes, such as performing analyses and designing an office space, must take place together with "soft activities."

Employees must be prepared for the new style of working and learn how to use and share the new spaces.

In turn, management must acquire the knowledge and skills that help to manage staff in the new work environment effectively. It is advisable to carry out workplace change activities with the help of external advisors who specialize in this field and implement these changes on a daily basis.

RULES FOR WORKING WITH SHARED WORKPLACES

Working with shared workplaces requires the introduction of a number of important rules in the workplace in advance. It is not possible to use fixed telephones within an office that is designed to share workplaces. After all, no employees are assigned to specific areas.

If there is a need, for example, in a department where many shared documents are used, it is necessary to equip the office with cupboards or an archive. This way, all documents are available for all employees who share a workplace.

SOUND INSULATION IN THE OFFICE: WORK IN PEACE

Do you sometimes feel that the office is more like a busy station hall? That is not surprising: people are in a hurry, having loud conversations, and there are telephones that keep ringing; it is a busy place on the work floor every day.

Noise pollution in the office is a problem that most employees have to deal with.

Intense, unwanted noises — and everything else that we call noise — contribute significantly to a decrease in effectiveness and loss of concentration. To reduce the negative impact of noise pollution in the office, we must understand the source of the nuisance and exclude it where possible.

One effective way to create a good acoustic environment is to choose suitable furniture and smart solutions for sound insulation around the office.

OUR COLLEAGUES ARE THE BIGGEST DISTRACTION

The main source of unwanted noises in the office? Our colleagues. The human voice is one of the most disruptive sounds. We are naturally curious and easily distracted. We eavesdrop on other people and, often unknowingly, try to analyze the content of conversations in the background, especially if we can only hear one end of them.

So because the sound insulation in the office is not in order, when we hear colleagues on the other side of a room make a phone call, the chance of working effectively is reduced.

WHERE DOES ALL THAT NOISE COME FROM?

Ringing telephones, buzzing printers, shrieking paper shredders, rattling keyboards, clicking mouse buttons, thumping footsteps — in addition to the voices of other people, there is a wide range of other unwanted sounds within an office.

The space in which we work often contributes to an increase in the noise level caused by these sources. An office design without sound-absorbing finishes and good sound insulation to isolate the sounds coming from workstations and office equipment gives these unwanted sounds all the space to spread freely. This leads to a cacophony of office noise, with an annoying reverb as a result.

NOT HAVING TO HEAR EVERYTHING

In a large room where working devices do not cause distraction, the noise level caused by conversations between colleagues is higher. Because conversations are not masked by other office sounds, conversations can be understood word for word, even in the most remote corners of the office.

This situation is a major obstacle to others' concentration, making them unable to perform their duties properly due to this distraction.

A possible solution is to install a sound-masking system. This network of small speakers masks sounds by producing an inconspicuous background noise, such as music, a soft noise (like white noise), or the calming sounds of the sea or the forest.

Would you rather not experiment with a masking system? Then there are alternative, suitable interior solutions to improve acoustic conditions and sound insulation in the office.

UNWANTED SOUNDS DISAPPEAR WITH SOUND INSULATION

After an analysis of the existing or potential noise sources, the next phase follows: reducing unwanted noises in the office.

Suitable furniture and interior solutions can help with this.

Every piece of furniture influences the way in which a sound wave spreads in a room, whether negative or positive. In an office with few sound-absorbing surfaces, the reverberation and noise can be reduced by installing benches or modular systems with properties that absorb sound better.

Please note: the principle "the more, the better" does not apply in the case of sound insulation in the office.

LISTEN TO A SPECIALIST

The excessive use of sound-absorbing materials and furniture ensures that an office is transformed into an "empty" environment, in which conversations between colleagues in the neighborhood are more clearly audible and thus provide more distraction.

That is why manufacturers, even in the case of sound-absorbing furniture, provide targeted advice with regard to using the right products, and the number of furniture required to meet the desired noise-reducing standards within an office is based on the current situation.

A sound expert can also indicate where the furniture should be placed in order to optimally fulfill their function and to improve sound insulation in the office.

A SUCCESSFUL SETUP FOR GOOD OFFICE ACOUSTICS

A storage cabinet is a good example of furniture that improves office acoustics. Do you choose this alternative to divide your office space? Keep in mind that the cabinets are at least 1.4 meters high; this can have a beneficial effect on limiting sound transmission between adjacent desks.

Once you install blinds and perforated doors, an office cabinet turns into a sound-absorbing "trap" that often works more effectively than upholstered furniture and wall or desk screens filled with a porous material. Especially in rooms without an acoustically lowered ceiling, this solution is important for creating pleasant office acoustics and sound insulation that works well.

SOUND SCREENS AND SOUND PANELS AT "EAR LEVEL"

Another important aspect that mainly concerns offices with open spaces is limiting the spread of direct noises between workspaces.

Desk panels and sound screens lend themselves well to this, but the correct height is also essential here.

Sound panels shorter than 1.2 meters offer privacy, but the sound has free reign. That is because the average "ear height" when sitting behind a desk is 1.2 meters. The sound wave squeezes itself around the edges of the sound panels and creeps into any free space, resulting in sounds being amplified rather than muted. In this case, keep a minimum height of 1.4 meters when placing a sound screen.

FLOOR COVERING MAKES A WORLD OF DIFFERENCE

Do you choose to designate informal meeting places in an open space? Consider dividing this zone with modular seating systems with a tall separation. This isolates the sounds of conversations from the inside.

This type of upholstered furniture can also have good sound-absorbing properties. Just like any piece of furniture, every surface in an office influences the quality of the acoustics. This also applies to the floor. Floor coverings suppress the unwanted sound of elevators, moving chairs, and clicking heels and can also contribute to the reduction of reverberation in a room. No matter how thin or thick, compared to smooth concrete, floor coverings make a world of difference when it comes to sound insulation.

BECOME A SOUND WAVE CONDUCTOR YOURSELF

As you read, there are various solutions to improve the acoustics within a room. And we haven't even talked about the sound panels mounted on the wall and on the ceiling, which you now find more often in offices.

The noise barriers — often filled with foam, glass wool, or felt and covered with a breathable covering — direct the sound waves through the room.

Are the panels standing or hanging in the right place? If so, then the negative noises that bounce through the surface of the wall or ceiling through the office are reduced. The reverberation also decreases due to sound insulation, making a room quieter.

Freestanding elements in the form of columns or corner panels also appear on the market for acoustic products. By placing these products in smaller meeting rooms, you prevent echoes or unpleasant rumbling from disrupting a good conversation. And that is certainly not merely a luxury in the case of important telephone meetings with several people.

A DEAFENING OASIS OF PEACE

With all this information, we know that it is important to be aware that the right office layout and conscious choices for furniture, materials, and finishes have a significant impact on office acoustics and noise levels in the workplace. The use of just a few of these elements and adjustments to the furniture arrangement can already help in controlling noise and improving sound insulation in the office. Call in the help of an acoustics expert who specializes in office environments to find the best solution for noise pollution.

The result: a deafening oasis of peace.

THE PLACE OF WOOD IN THE MODERN INTERIOR

Wood is like a movie that you have watched many times or the persistent urge for that "good ol'" CD or record player. Although we have an increasing number of modern materials at our disposal, such as MDF, linoleum, or melamine, we continue to opt for wood in our interior.

Where does this affection come from? Why do designers remain passionate about the use of wood in their projects? And what is the reason for that passion?

PLASTIC DOES NOT KEEP WOOD OUT

Wood is a material that has been used for centuries for making large and small structures, everyday objects, and — especially — furniture. The moment people were given the opportunity to process wood, the material became the basic design of all interiors.

In the last few decades, wood has been embroiled with plastic in an intense battle. Plastic is available on a large scale, cheap, and easy to process and maintain. Despite all these characteristics, however, it is wood— even now — that has remained the basic material for designers. Because apart from the functionality, wood enriches every interior and makes this material a real eye-catcher.

A PRODUCT THAT GROWS BEAUTIFULLY

How is it possible that wood has been and will continue to be the material that we use for making furniture for centuries?

And what makes wood so unique that it keeps returning to trends in interior design like a boomerang?

First of all, wood is an extremely elegant material that gives every room timeless charm. Moreover, wood is solid, durable, and can take a beating.

Even if wood ages, the material does so with dignity. The passage of time refines wooden elements and structures.

UNIVERSAL, VERSATILE, AND TIMELESSLY POPULAR

Thanks to the universal and versatile nature,

wood can easily be combined with other materials and fabrics. This is clearly visible within modern design trends and the approach of designers. Interior specialists are increasingly opting for the combination of wood with non-obvious materials, such as fabric patterns, linoleum, or laminate.

As a result, they create beautiful, unique, and, above all, surprising design combinations.

WOOD AND MODERN OFFICE TRENDS

Research has shown that the presence of natural materials in the office helps to reduce stress and has a positive influence on the well-being of employees.

History shows that designers and architects work passionately with wood, but contemporary interior trends encourage the use of this natural material.

This proves, once again, that wood is an ideal solution and a perfect addition to many different office spaces.

Precisely for this reason, within the biophilic design trend — intended to make people live and work in harmony with nature — placing large amounts of greenery and natural materials within a space is one of the most important principles.

Employers nowadays often opt for this style of interior design, because wooden elements contribute to the satisfaction of employees, stimulate their creativity, and increase work efficiency.

THE INDISPENSABLE INGREDIENT FOR THE HOME OFFICE

At home, in our apartments and homes, we make full use of wood. Wooden furniture and accessories add warmth and coziness to our private domains. This knowledge and experience mean that interior designers nowadays increasingly want to convey this specific, homely atmosphere in office environments. This is achieved by, among other things, the use of different fabrics, but wooden furniture cannot be missing. Wood can be used in many ways in the "home office," from completely solid wood structures to the subtle use of wooden elements, such as table legs, desktops, and handles.

11. A SCANDINAVIAN INTERIOR FOR YOUR OFFICE

Have you ever had a cup of coffee in an office with a Scandinavian-style interior?

If not, you now get the chance to imagine yourself in such a relaxing environment. We'll take you along in this challenging interior trend and show you why these specific colors and furniture choices fit perfectly within your organization.

FUNCTIONAL, CLEAR, OPEN, AND WARM

The characteristic features of the Scandinavian style are functionality, bright and open spaces, and warm colors, all of which have a positive effect on the well-being of people.

Lighting plays an important role. Even if there is insufficient access to natural light, it can be provided by suitable and friendly artificial lighting.

Perfectly planned zones ensure comfortable working in different circumstances, from team meetings at high tables to spaces where people can work with great concentration.

The Scandinavian interior is, above all, practical; there is no coincidence; every solution is justified within the Scandinavian design.

HYGGE: THE NEW RECIPE FOR HAPPINESS AND COZINESS

This happiness recipe for creating a warm, friendly atmosphere — for living and working — is based on an apparently simple design.

Natural materials and earth tones such as beige, brown, and gray are perfectly mixed with pastels such as pink or blue, which evoke the colors of the sky. Offices with a Scandinavian design that embraces and radiates *"hygge"* are warm and cozy. Soft and natural fabrics dominate; cotton, felted wool, expressive fabrics, and accessories such as pillows, ottomans, and blankets evoke a cozy, homely feeling.

NATURE DOMINATES IN A SCANDINAVIAN INTERIOR

The main purpose of an office environment is, of course, not just to give a pleasant feeling. Work is also needed.

Completely in line with the idea behind a Scandinavian interior, the emphasis in this style is on a comfortable workplace.

Classic open spaces where desks are arranged in rows make way for loose, raised desks, cupboards, and storage furniture or acoustic walls. Height-adjustable desks and chairs — with a fabric backrest — are an absolute must-have in Scandinavian design.

Natural colors and materials, such as wood, dominate the interior. We also see metal, for example, used for chairs or table legs.

The metal is often painted black or white; shiny chrome is definitely not in place in this establishment. Monochromatic, pure color combinations are also increasingly being used.

A POSITIVE EFFECT ON WELL-BEING, WORK AND REST

The combination of subtle, harmonious beauty and functionality creates an atmosphere that has a positive effect on the well-being of users, which is conducive to both work and rest.

All these pieces of a Scandinavian interior, although fashionable and widely used in interior design in recent years, are still dominant interior trends, also in the design of office spaces.

Happiness at Work

Monday: pamper day

Start the week with breakfast among colleagues, without talking about work: the atmosphere is good right away! Everyone goes back to work with a smile.

More light, please!

People need light, especially in the winter. But too high a light intensity we experience as being aggressive. Give preference to soft light and avoid cold neon light to create a cozy atmosphere.

A short break

Occasionally, a break is essential to perform optimally for the rest of the time, but in Scandinavia, they go even further: short gym sessions, relaxing massages, micro-siestas somewhere comfortable...

Flexible timetable

In Denmark, a working week lasts a maximum of 32.5 hours, but flexibility is more important there. Every employee optimally organizes their working time.

Trusting people

No gossip, jealousy, or humiliating criticism... Everyone has the right to make mistakes, and respecting others is the basic rule.

12. HOW TO CREATE A COZY HOME

1. Use soft and inviting seating

The alpha-omega of rest hygiene is that you can sit down and relax your legs, whether on a soft sofa or in the #1 cozy furniture: a good armchair. In fact, if you are looking to create comfort, consider the seating areas of your home as small "places to stay" that should be nice to stay in. Here are a number of different elements that you can read more about on these pages. But let's start with the furniture: choose large and few furniture instead of many small ones. It creates wholeness rather than a mess. However, it is a good idea to mix different types of furniture and work with contrasts in soft shapes and lines — this will make the interior more personal than if you decorate with a large furniture group where everything has the same expression. One tip is also to pull the furniture away from the wall — because even if you don't think it will, it will make the space seem bigger.

2. Group your cases

It creates a good rhythm and tranquility in the interior to group everything from furniture to pictures on the wall. It is classic that for fear of not having enough things, plants, and images are spread in a large area. But it is much better to group what you have. For example, gather the green plants in a cluster, hang the pictures in groups on the wall, and put small things on the shelf together in small compositions with a little air around.

3. Invest in plaids

Coziness is a feeling, and that feeling requires something soft to wrap yourself in. Therefore, coziness is almost inextricably linked to a plaid or more. But teas are not only nice because they are warm. They also soften the interior, add tactility, and make a space more welcoming. A sofa can appear naked and cold without a plaid over the armrest.

4. Complete the color palette

Whatever color palette you fall in love with, it is important to complete it — both in relation to the base of the room, i.e., the color of the walls and ceilings, as well as in the fabrics you choose to decorate. If you are brave — and fashionable — then paint the ceiling and wall the same color.

This way, you create a space with a cozy cave atmosphere. But less can do it too. The monochrome expression can also be created by simply painting walls, moldings, and frames in the same shade. Next, you will have to expand in your choice of textiles, which should move within roughly the same color palette. If you are unsure which palette to choose, stick to natural materials.

5. Be personal

A cozy home reflects the people who live there. Heritage, memories, and objects from a special time help create a safe and relaxing setting, so use it in your décor instead of hiding it away. That is not to say that you necessarily only have cases with personal value on the shelves, but also remember that there is no other place where you are more yourself than when you are at home. It also creates a great atmosphere if you can actually see that someone lives in your home. That the book you are reading is allowed to advance. That the pillow does not necessarily fit perfectly on the sofa. Your guests will certainly also feel more at home if there is a relaxed atmosphere when they visit you.

6. Let the light set the mood

Light is an important mood creator, but it is an art to make electric light seem cozy and natural. The main rule is to combine cold and warm light, and you can, for example, work with cold LED light in combination with a warm incandescent bulb, so that you give yourself different lighting options for different situations. For that, dimmable light is also (and still is) a good solution. The location of lights and lamps is also very central, and it is good to have both floor and reading lighting as well as in the ceiling.

However, limit the use of spotlight — it gradually overlaps with the past to light up an entire room. Today, think about what you want to light up. If a wall is illuminated, it may shed light on the rest of the room, and you may not need to illuminate the others. In fact, coziness requires as much shade as light — and the nuances make your home more interesting.

7. Choose the right materials

The materials are very important for the sensory impressions you get when you enter a room. And if there is something that pleases, it is soft materials and fabric, but it is also a good idea to combine matte surfaces with glossy objects such as brass candlesticks. It gives the fire a cozy and vibrant glare that contributes to the atmosphere.

8. A matte base can contribute to a warm atmosphere

Matte paint on the wall and ceiling is currently really popular, but the matte expression is more than just a trend. Matte walls create a denser atmosphere in the room, which is calmer, warmer, and softer. For example, if your wall is painted with a lime paint, you also get more play in the wall, which contributes to a warm atmosphere, while a blank wall mirrors the room without adding more.

The same goes for glossy floors that can seem cold.

How to Get the Most Out of a Large

Living Room

DECORATING IN GROUPS

To take advantage of the many square meters in the living room, it is a good idea to arrange things in groups to create coherence in the room, but first, you should consider what primary functions the living room should have. For example, a large living room can be divided into a dining area, sofa area, and a cozy corner. It may be difficult to create groups in the living room if there is no connection between the different departments. By placing a rug in the dining area and in the sofa group, the rug marks the areas of the interior that will create a sense of overview and create peace of mind.

SOFA AREA

The sofa group should be fully utilized when you have plenty of room in the living room. Whether you want a large sofa with a chaise longue, a three-seater sofa, or a combination of several sofas, you can easily make the sofa area cozy.

Although we often place sofas up against the wall for more space, it may be a good idea to place the sofa area away from the walls and out to the center if you have a lot of floor space. That way, the living room will not seem empty and, at the same time, create a cozy atmosphere.

Another benefit of moving the couches away from the walls is that you can utilize the walls for shelves, ladders, and posters, just to name a few examples.

COZY CORNER

If you have a large living room, there should certainly be a small cozy nook where you can enjoy your own company. The small corner of comfort should delight your eye, making you want to sit down. Therefore, it is optimal to place a stylish lounge chair that adds a unique look to the room.

Next to the lounge chair, you can put a nice side table, for example, in marble that has a nice play in the countertop. After that, you should acquire a lamp according to what you want to use the cozy nook for.

HIGH CEILINGS?

If you have high ceilings, you can place delicious and large ceiling lamps over the dining table. That way, you can create lots of volume with your lighting and get a nice eye-catcher in the living room.

In general, a large living room can seem cold and open, so it is a good idea to decorate the living room with a lot of lighting — both focused light and cozy light. Therefore, use large floor lamps and ceiling lamps, which, together with the cozy lighting from the table lamps and wall lamps, create a cozy living room.

GET COZY

Coziness can be created through lighting, but you can also make the living room more homey with pillows, blankets, and blankets. In addition, both small and large plants can help make the living room more welcoming and stylish. Place plants on the floor, coffee table, bookcase, and many other places that, together with candles, can do wonders in a large living room.

If you still have plenty of room in your living room, you can create an exclusive look by placing a daybed in the living room. Move the daybed away from the walls and let it be decorative on its own — you might even decorate it with fur.

Sofa

The most essential in the living room is the sofa. This is where you sit and lie, depending on whether you should lie down and relax or sit and have a conversation. There are countless different sofas, which is why it is important to find the one that fits right into your décor and space.

You can choose a two-seater sofa/loveseat. Such a sofa is particularly suitable for small rooms, but there is not much room on them, however. If you want to upgrade, there is also a triple sofa. Here, a little more space is needed, but there is also more space in it. If you have plenty of room, you can choose to get a U-sofa or a sofa with a chaise lounge. These are especially beneficial if you are many in the family or usually have many guests. There is a myriad of models and combinations of the different sofas, which is why it is also a big decision to make, as you would prefer to use the sofa you find in the first attempt.

The sofa is the middle of the room. Whether it's colorful or with neutral colors, like beige and gray, depends on your personal taste. The most important thing is that you can sit comfortably and also be able to lie down.

Lighting

Indirect lighting with LED strips, conventional floor lamps, table lamps, or sconces ensures a pleasant and comfortable atmosphere. Dimmed light gives a sense of peace and security.

It is especially nice and romantic, as usual in Sweden, to put lights on the windows. It creates light from the position where daylight penetrates during the day, which eases the need for sunlight. Moreover, this light is a sign of safety. Previously, when the streets were not lit, the small light sources in the windows showed the way home.

Brands

Nothing can give a room more warmth and coziness than flames in the fireplace. The classic wood stove and the eco-friendly alternative bioethanol fireplace are both a decorative highlight. One advantage of the bioethanol stove is that it does not require a flue gas chimney.

Colors

Play with the colors. Each color has a special effect on people: orange radiates warmth, yellow brings sunshine to mind, red has a stimulating effect, green gives harmony, blue has a harmonizing and relaxing effect, violet promotes inner balance, and white gives clarity. You can combine the colors endlessly and change quickly, depending on the mind, with a new cushion cover or similar accessory.

Accessories

The finishing touches to the interior of the room are provided by accessories. You can give your special personal note to the room: pictures of loved ones or your travels, original paintings, reproductions, or posters on the wall. And the indispensable candles that bring warmth and comfort to the room quickly.

Furnished with blankets, blankets, and pillows

Soft fabrics in fine patterns and colors can help boost the mood of your living room. If you have wood flooring in the living room, it is a good idea to find a large carpet to throw under the coffee table. Buy a nice plaid blanket for the couch that you and your family can wrap up in. And find soft pillows that match the colors that are otherwise in the living room. Try finding pillows that will make the couch even more alluring.

Cozy with light and heat

Even when it's not winter, but especially when it's winter, lit candles can really make a difference to the atmosphere of a room. Flickering fireplace lights in any nook provide warmth to a room, and if you have a fireplace or woodstove, you can also just turn on the heat when you need a little extra comfort.

Listen

Coziness is a feeling in the body, and it is a treat for all senses. Therefore, do not settle for a soft blanket you can feel or candles you can see but put sound in the living room as well. Whether it's an acoustic indie band you love or the *Mads and Monopoly* podcast, it will make a positive difference with the mood of the room.

Let nature inspire

Another way to create comfort in the living room is to let nature come with its input. Find your favorite plants and place them on the windowsill or wherever appropriate. It also helps to create peace of mind when watering and nourishing the plants.

Put away all the work

If you sometimes work from home or have papers or similar to ones from work, put it away. When you are in the living room and relaxing, the last thing you need to think about is your work.

Make the living room a mobile-free zone

We love our smartphones, but there is also a lot of stress associated with those small devices. For example, you probably have your work email on the phone, and if you don't want to check it while you relax, put the phone away. It's also an incredibly effective way to create presence if you set such a rule for your living room. Instead, play a board game with your family or focus 100% on the movie you were watching. Facebook,

Instagram, and random Google searches can wait.

13. HYGGE LIVING: TIPS FOR DANISH COZINESS IN THE LIVING ROOM

Plan Your Living Room: How to Create a Healthy Living Environment

A cozy ambiance is important to make you feel at home. In order to create a healthy living environment, various measures are necessary. Combine your optical living room ideas with practical, because your new living room should, of course, be not only beautiful, but also functional. By choosing the right furniture and materials, you also ensure a completely healthy indoor climate.

Define a concept for interior design.

Before you start with the actual planning of the room and go into colors or lighting, it is advisable to define a design concept. This must respond to your individual needs and, of course, match the spatial conditions. Try to answer the following questions for yourself:

- Is a dining table in the living room

important to me?

- How much storage space do I need in the living room?
- How should the style be — classic, modern, elegant?
- Should the living room furniture be rather simple or extravagant?
- Do I want to continue using my existing furniture?
- Should the living space be designed to open or closed?

A blanket statement for the questions cannot be made, because these are essential for each room and adapted to your individual needs and desires. If, for example, you would like to continue to use existing furniture, you should adapt the colors and further furnishings to the existing living room design. But not only furniture, decoration, and materials play a role; you also have to include the light. In a rather dark room, for example, the furnishings should be brighter, while a light-flooded room can tolerate dark and gaudy furniture.

Virtual tools help set up the living room

Have you already thought about the design and put your living room ideas on paper? Then you can start with the final planning. The easiest way to do this is with virtual tools that are offered by numerous providers for free. First, enter the exact dimensions of the room according to the floor plan, and you can virtually insert living room lamps, furniture, and accessories into the room. Pay attention to the height of the room, because especially in small living rooms, you can create a lot of space with high closets.

The advantage of virtual programs is that you have the chosen furnishings and living room colors right in front of your eyes — usually even in 3D — and can see immediately whether the chosen colors and patterns harmonize with each other.

Cozy light and candlelight

Winter in Denmark is long and dark. Brightness is, therefore, an important topic for the Danes, even within their own four walls. For more *hygge* in the living room, many windows, many lights, and candles are important. By the way, did you know that the Danes consume about two to three times as many candles a year as Germans do? In the evening or on particularly gloomy winter days, harmonic light sources play an important role.

Warm, natural tones

From cream to white: Scandinavians love bright tones that reflect the light in all corners. White walls and furniture also brighten the room if it does not have quite that many windows. In addition to a light couch or a beige armchair, white lacquered floorboards are often to be found in Scandinavian living rooms. However, the whole thing does not seem sterile, as it is played with soft textiles and warm materials — an important element for the *hyggelig* setting.

Grounding natural materials

Speaking of warm materials: *hygge* also has a close relationship to nature, so wood is an integral part of a cozy, *hygge*-style living room. A paneled ceiling, wall, or both is not stale, but typically Scandinavian. Modern options are bright wainscoting of pine, spruce, or birch, but also teak for mid-century flair or old, reclaimed wood for a rougher look — these options all fit well into the *hyggelig* living room. Even on the floor, you will rarely find cold tiles in

Scandinavian living rooms. Floorboards or parquet not only keep feet warm, but they also give the room a homely atmosphere.

Pattern: anything but monotonous

Clear lines like the northern lights are great for the accessories. Cushions, blankets, or carpets with graphic prints are popular. Rhombuses and triangles on textiles fit as well as vases with circular imprints. A well-known example, where sometimes it is colorful and wild, is the prints of Marimekko from Finland. The effect is similar to simple color accents: the patterns bring movement into your living room and just make you happy. As always, accessories are the fastest way to get a new effect without having to redesign the entire living room.

More coziness with skins and blankets

Feel good, make yourself comfortable — cozy wool blankets or soft sheepskin (as a throw on your favorite chair) are the epitome of *hygge*, because they stand for coziness, naturalness, and warmth like no other. If you do not want to bring real fur into your living room, look for beautiful faux fur and thick-knit or wool blankets. The main thing is cuddliness.

Practical helpers for convivial evenings

The *hygge* feeling for life is easy and always somewhat improvised. This is because it is more about a good mood and less about the outer shape. This also applies to the *hyggelig* device, which relies in the sitting room rather on a group of mobile side tables, than on the one heavy and thus immovable coffee table. On the one hand, this makes the room seem airier; on the other hand, such a handy side table is pushed or carried quickly from A to B — very practical for a spontaneous visit by friends.

14. STEPS TO A COZY BEDROOM

1. Mood maker around the bed

Dream, cuddle, and feel good: a special armchair or a small couch can make any room cozy and invites you to daydream and browse. If the room is also to be used for fitness or for work, a screen or a shelf can obstruct the view of the equipment.

2. Selected favorites

A lavishly furnished room is rarely cozy. Better to place in the room only a few but selective, favorite pieces that harmonize in form and color: a small table near the reading chair, a single dresser, or beautiful pictures.

3. Sleep with Feng Shui

If you want to relax, according to the Feng Shui doctrine, avoid protruding edges and corners.

Place the bed in visual contact with the door, but not too close. Likewise, sleeping between door and window (or right in front of the window) hinders the harmonic *Chi* energy flow.

4. Play with the proportions of the room

A small room looks bigger if you use light shades or wallpapers and, if possible, do without patterns; whereas, strong colors make a large room look smaller. In low rooms, it is advisable to make the ceiling a few shades lighter, so the room visually gains in height. Such an effect can also be achieved with a ceiling washer.

5. Bright colors give width

White looks generous, but also cool and clinical. Use instead light pastel colors or warm terracotta tones. Look for a uniform color family for different shades. Otherwise, the room will quickly become too restless.

6. Storage in small rooms

So that wardrobes do not act as optical troublemakers, pay attention to uniform colors on the walls of the room and door fronts. For example, furniture of a considerable volume also visually fades into the background. The same measure helps to hide the size of built-in wardrobes.

7. The right lighting

The lighting concept ideally has three main focuses: (1) general (ceiling) lighting with bright and functional light, (2) a bedside reading lamp, plus (3) mood lighting for relaxing, cuddling, or daydreaming. For mood lighting, use several light sources or individual spots with a warm light tone, which discreetly stage individual areas or a photo. It is advantageous if all lamps can be operated at the door and also from the bed.

8. Reading pleasure in bed

For cozy browsing, we recommend a table or wall lamp with swiveling reflectors. In this way, the light can be directed exactly where it is needed without dazzling the bed partner.

9. Curtains and fabrics bring living atmosphere

The colors and patterns of curtains, blinds, or bedspreads are ideally also matched with the wall and floor. Tone-on-tone harmony has a favorable effect on well-being and promotes relaxation.

10. Clear conditions on the walls

The wall surfaces are designed to be calm and uniform. So do not clutter too many small pictures, but consider a clear order principle for attachment. So the room is quieter, by the way, the pictures show much better.

How to decorate the most beautiful bedroom

Bedroom

In the civilized world, about 40% of the population complains that they cannot find a good night's sleep. Another 40% say they suffer from a severe sleep disorder. Would you have thought that? These values are frightening, given how important a restful sleep is to our health.

Tip: A restful sleep environment and sleeping space are essential for healthy sleep

The reasons for bad sleep can be manifold. What is certain, however, is that the sleeping environment and the sleeping area itself are an integral part of it. Since we spend about a third of our lives sleeping, with the right bedding and sleeping environment, we should lay the foundation for healthy sleep. Topics covered here include orthopedic sleep systems, choosing the right pillow or bedding materials, mattress and slatted base.

Some examples: If you wake up every morning with neck pain, too high, or too low, your pillow could be at fault. If you frequently sweat during your sleep, you should consider the choice of bedding materials. Too hard or too soft mattresses are often reflected through back pain.

However, not only does the sleeping place itself contribute significantly to our sleep, but the environment should also be perfectly matched to it. In addition to the room and air climate, this includes choosing the right colors as well as the furniture and technical equipment used. You will be amazed at the ease with which you will be able to improve your sleep after a short time.

The bedroom is one of the most important rooms in the home because you spend so many hours in it.

Staying in the bedroom should be nice, and you should wake up well-rested every single day, which is why you get a guide here on how to decorate a really nice bedroom.

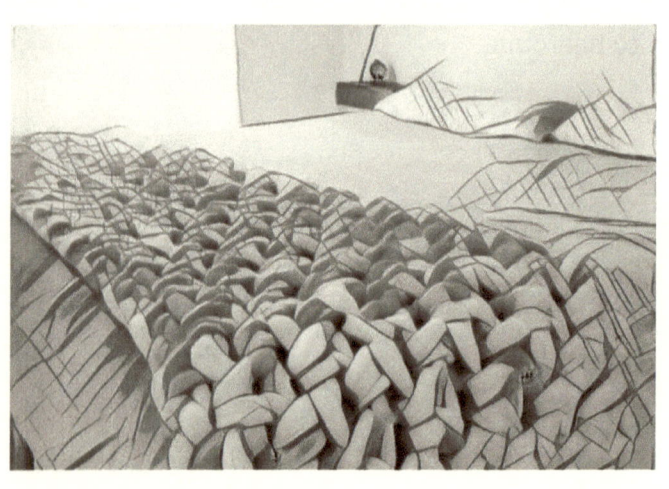

Buy a good bed

The bed is unmistakably the most important piece of furniture in the bedroom, and perhaps even throughout the house.

Beds that do not fit one's body or one's sleep style can actually have fatal consequences for both one's health and one's mental faculties, and so it is really good to invest some money in a comfy bed so you can sleep well every single night.

Choose functional furniture

There should be tidiness in the bedroom, and to make it easier to keep in order, it is smart to buy functional furniture.

Of course, it does not matter if they are at the same time decorative and can help make the bedroom inviting and personal, but the most important thing is that they are functional, so you can easily keep the mess away from the bedroom.

If there is a mess in the bedroom, it will be difficult to relax and disconnect, and therefore, it is worthwhile to make an effort to easily keep track of clothes and other things that could otherwise be lying around and floating around.

Create lots of personalities

The bedroom should very much be tidy and practical, but at the same time, there should also be lots of personality in it.

You relax best if you are in a safe environment, and you will undoubtedly feel comfortable and more at home in your bedroom if there are nice holiday pictures on the wall or your favorite flower in the window sill.

Even if you spend many hours sleeping, there should also be a room just to sit down with a book in the evening and relax, and that makes you feel like if a personal and cozy atmosphere has been created in the bedroom.

So please do not: storeroom and office

On the other hand, you should always avoid the following mistakes when setting up your bedroom: it should not be misused as a storage room where vacuum cleaners, moving boxes, and co. find their place next to or near the bed. This not only looks uncomfortable but it is also at the expense of a healthy night's sleep. In the same way, the use of the bedroom as a study (which is sometimes unavoidable in smaller apartments) does not contribute to undisturbed sleep.

Setting up your bedroom: It's all about the furniture

The central piece of furniture in the bedroom should be the bed instead. Ideally, it is large enough and tailored to your personal sleeping habits. It is also important that the slatted frame and mattress fit together: foam mattresses, for example, only lie correctly if there is a small gap between the slats. The material of the bed frame, on the other hand, is not crucial — more important is the location. Place the bed in a quiet corner, so you do not sleep with the door at your back.

How to hide your work
So that the bedroom does not seem overloaded, it is worth a closed wall unit, behind which is plenty of storage space for clothing. Even disturbing items such as an ironing board or laundry basket should find their place here unless you have a storage room. The less "objects of work" distract you from sleep, the more easily you will find peace in your personal retreat. If you are so limited on space that you can only work in the bedroom, you can hide the computer and desk with a screen or curtain, so they are invisible from the bed.

Light and color for more coziness

Whether you completely darken the bedroom with blinds or light-tight curtains, or whether you want light curtains or blinds, depends on your individual sleep type. In terms of room lighting, the light should not be too bright. Dimmable lights with warm white LEDs and unobtrusive, indirect light create a cozy, sleep-inducing mood.

When it comes to color design, it pays to restrain the bedroom furniture. A wall in bright pink or blood red does not necessarily guarantee undisturbed sleep. Intense colors should be combined rather than details with lots of white to create visual peace. Otherwise, pastel colors or subtle blues or earth tones provide a calming room atmosphere.

The bed and bedside table

Of course, the bed must be perfect; it dominates the room and accompanies you practically a lifetime. In it, you draw new strength and energy so that you are fit again the next day. So the quality and above all, the selection of a suitable mattress have a big influence on our sleep.

Since everyone needs different criteria for healthy sleep, it is certainly best to try out mattresses on the spot. You should pay particular attention to the bed, which is the focal point of every bedroom.

Create a feel-good oasis; you need to fall asleep in bed, lie well, and wake up refreshed. The model must be well chosen, but there are enough beds, and a model will probably suit your taste.

A bedside table just belongs to the bed, where else with the alarm clock, the lamp, or the book. Bedside tables are available in all variants, in every style and with or without drawers. Some models have doors, and others have a second floor to accommodate your bed-reading.

Of course, a good piece must fit the bed; the height of the piece of furniture is very important. You can set accents with the side tables and choose these tables in bright colors. But also very discreet bedside tables are on offer, made of wood or metal.

Bedroom setup — 5 cozy furnishing ideas

1. The wardrobe

Of course, a wardrobe in the bedroom should not be missing; it keeps all your favorite things safely and professionally. It creates order and offers each garment its own place. Meanwhile, you will also find plenty of organizers, with which you create additional storage space in your wardrobe.

You will quickly find the right model, with wardrobes in classic, romantic, and very modern forms. This is a piece of furniture you cannot do without; this is the only way your bedroom will always looks neat.

2. Wall design and lighting

It is a very important aspect when setting up the bedroom; you create the right atmosphere in the room with colors, wallpapers, and pictures. Wall colors already set the mood in the bedroom: gray, blue, green, or beige is perfect. These hues create a bright and rather cool atmosphere.

With white, you cannot go wrong; this color is never intrusive and is particularly suitable when you set up a small bedroom. Wallpaper is also a very good idea, but the pattern should not be too colorful.

Better is a subtle wallpaper that underlines the style of the bedroom. You can also attach a photo wallpaper with your favorite motif to a wall. Modern photo wallpapers are currently very trendy.

Without light, it does not work; at night and on cloudy, rainy days, you need enough light. But the lighting does not always have to be bright; indirect lighting must not be missing when you need it.

Of course, bedside lamps and floor lamps also create a romantic mood; the lights need only be well-distributed. Lamps for the bedroom are available in all sizes and in various variants, whether small or large, cute and playful, or simply in clear areas.

3. The bed linens

What would a bed without bedding be? You cannot imagine that, because only the bedding can make a bedroom perfect. You should not save on the bedding; it should be soft and cuddly and allow for comfortable sleep.

The fabrics can vary; in the summer light, cotton fabrics are just right, and in the cold season, beaver bedding is great. Sheets are available in all sorts of colors, and with different patterns, you can transform the bedroom in no time and achieve a completely different look.

Pillows can be put together as you like, but you should not miss them on the bed or on the rocking chair. They are a real eye-catcher, and at the same time, they provide a soft underlay for your head. You can play quietly with the colors and patterns; you cannot go wrong here.

4. Carpets and curtains

Carpets also play a big role in the design of the bedroom. On the one hand, they keep the feet nice and warm, but at the same time, they are a successful decoration. Complete the look of the bedroom with rugs; they are available in any size, with any pattern, and in all price ranges. If you like it cuddly and cozy, then lambskin or high pile carpets are ideal.

Curtains are available in all variants today. Translucent curtains are simply designed for necessary privacy. These fabrics let light through, while you go completely unobserved. In order to completely darken the room, however, an opaque curtain must be installed. These curtains are made of thicker fabric; with them, you will sleep wonderfully, even without a roller blind.

Curtains are also available in a large selection; countless colors and patterns are available. However, bright and simple colors emphasize the atmosphere in your bedroom.

5. Decorations for the bedroom

Décor items for the bedroom are really plentiful. Make your bedroom appear again and again in a different light, the decoration you can always redesign. Candles should definitely be there; they conjure a very romantic mood.

Do you have a green thumb? Then indoor plants give a very special charm. Colored flowerpots or colorful vases set accents and are the right splash of color. Wall decoration is an eye-catcher anyway; beautiful and imaginative pictures make an impression.

How you design your bedroom is, of course, your business. However, furniture with a mix of materials is very modern, the times in which the wardrobe, bed, and bedside table always looked the same are over. Mix different styles and do not hold back on the materials.

Indispensable tips for your bedroom

It can be nice with seating

It is convenient to furnish the bedroom with a decorative chair. For example, it comes in handy when wearing shoes. Thus, the chair is not only a decorative object — but it is also practical as a bed and relief table and of course, as a chair.

Remember not to over-decor the bedroom

Remember that the bedroom, in particular, is the place where the body should find peace. You need to be able to get to different things, so you avoid distractions and frustration.

How do you use the space?

What do you use the bedroom for — besides for sleeping? Maybe you have an office, dressing table, TV, or a balcony? It is important that the bedroom is functionally furnished if you want to enjoy it in the long run. Therefore, customize your decor according to electrical outlets, windows, TV, and so on.

Partitions in the bedroom

Make your own partitions using shelves, curtains, clothes racks, or walls. This creates a brilliant space-in-the-room feel, and you also get several surfaces that you can set up your furniture. Like the bedroom, as you know, should invite for relaxation, it can be an especially good idea to divide the office and bed section.

Simplicity with cracks in!

A single bedroom doesn't have to be boring. Use your favorite colors as spices when decorating a single bedroom.

15. COZY KITCHEN SETUP — OVERSIZED KITCHEN DESIGNS

Kitchens are the heart of the house, and they gather friends and family all day long. All people feel more comfortable in larger apartments and want to have more space in the kitchen and in the other rooms. The kitchen is no exception. In over-sized kitchens, chairs, or stools can be found on the kitchen corner, and side seating in the adjoining dining area also offers more space.

Cleanliness is the first, smallest, and most important principle. Dirt can overshadow the grandeur of an expensive headset. And if there is no money for a new decoration or furniture, you should just try to clean the room. Everything that has dissolved over time, glue, mending, dusting, cleaning dishes, and plumbing work. Remove splashes of oil from the working wall, clean household appliances. Without this stage, the way to comfort is simply impossible. No curtains or furniture draw attention from the greasy tiles. Imagine how beautiful it is in the kitchen, where everything has its own place — where harmony and cleanliness prevail.

The Situation

Imagine how this can affect comfort. And let us think about it. What comfort does the majority have? Of course, with convenience. And now we look at the kitchen: where there is a lot of unnecessary furniture, it is difficult to open the fridge, you have to move the chair, close the window, etc. And is there any comfort here?

To get to the dining table, for example, you need to overcome several obstacles in the form of unnecessary stools or uncomfortable bar tables. Unfortunately, in many kitchens, there is still no comfort. So what is comfort you? Therefore, it is important to think about the layout of the main elements so that everything is functional, does not interfere with the freedom of movement in the room, and is really useful for the kitchen.

Lighting

Incorrectly installed light is the fatality of the kitchen. First, it is inconvenient to do chores, and secondly, incorrect lighting can make the room much smaller and more harmful. That's why it's important to get the lighting right for comfort. It is best not to be limited to a lighting device, such as a chandelier with a candle. The light should fall from several sides.

And for real comfort, it's best to use multiple contrasting sources. For example, illuminate the work area with brighter daylight bulbs and add some privacy to the dining area by using a soft yellowish floor lamp or a lampshade.

It is best to design the lighting so that the lamps work autonomously. If the devices are not interdependent, different lights can be installed. This means that the room gets different effects and creates a different atmosphere.

Color range

Visual comfort is no less important than physical comfort. The right combination of colors and shades is the first step in creating a harmonious interior. A cozy little kitchen in bright colors without mixing different colors looks spacious. Conversely, a spacious room with a wrongly chosen coloring, as well as an admixture of different tones, becomes a chest of drawers. Therefore, it is important to choose a color combination of harmoniously.

For small rooms, do not use more than two dark shades of the same color. General ideal option — use monotony or no more than 3 colors. Comfort is not too much combined with expression. So do not make the kitchen disco-style with rainbow stripes or bright, eye-catching colors.

Beautiful cozy kitchen in the photo — an example of soothing pastel tones that are best perceived by the eye from the point of view of the psychology of perception. In addition, the color requirements are not limited to the interior and different things.

First, the tonality of the goal is important. And if it is too dark, furniture and other objects should contrast a priori. Then all elements complement each other and can get a pretty nice picture.

Curtains

Who said that this is a completely unnecessary article for the kitchen? Imagine, many are pretty sure that there is no room for fabric curtains in the modern kitchen. And all this can be confirmed, provided there is no comfort. But no element gives as much warmth as the beautiful curtains on the windows. And many use roller blinds or roller blinds that are more functional. It is natural, but such products do not give any comfort.

Fresh flowers

They should not be too much; do not turn the kitchen, especially the small ones, in the greenhouse. However, you should put a small emphasis on the natural atmosphere. These elements emphasize the desired comfort perfectly.

Taking these simple rules into account, you can create a pleasant and harmonious feel-good atmosphere in any kitchen, regardless of its size.

The question of how to prepare a cozy kitchen worries the hosts of both the small and the spacious apartments. This chapter contains practical advice and recommendations on how to create a pleasant atmosphere in each of the smallest rooms.

Lighting

To create a cozy kitchen, you need to choose the right lighting. Everything depends on the stylistic orientation of the room. For the interior in the style of Hi-Tech, the lamps fit perfectly in cold colors, ie, in gray, black, or brown. The combination of the "gentle" warm yellow light of the chandelier and the cold, almost "sick" shadow of the headlights characterize the classic style of the style. If you want to make a cozy kitchen, you need to choose the right ceiling light that can be successfully hidden in special compartments under the ceiling. Posting can be both common and decorative in this case. The latter option is suitable when non-standard wires need to be focused.

Choice of furniture

How to make the kitchen comfortable with your own hands? For the furniture, the patterns of soft artificial leather or leather, which are painted dark brown, match the classic interior.

The shadow of the walls depends on the overall construction of the entire house. In this case, however, the combination of beige and peach blossoms can make the kitchen comfortable with your hands.

Bright world of textiles

Textile production also helps if you want to make the kitchen bright and comfortable. Beautiful curtains on the windows will fit bright gloves, various beautiful towels, aprons, and other trinkets. How do you make the kitchen cozy and beautiful? If it is monochrome and has no "eye-catching" colors in the design, it is better to use elements with bright colors. For example, green, purple, yellow, and other juicy tones are perfect.

The right dishes for the right kitchen

To make a cozy, beautiful kitchen, you should definitely buy a light dish, from which it is good to eat. Porcelain works very well, and in some cases, it can even be made of wood.

Needlework in the design of the room

To make a cozy kitchen with your own hands, you can show the imagination and the skill of manual work by connecting the tablecloth or the furniture covers independently. You can also hang a variety of paintings on the walls to create a cozy atmosphere.

Proper zoning

In answering the question "How to make a cozy little kitchen cozy?", one must remember decoration, such as the optical layout. This can be achieved by placing two different types of laminate or decorative coatings in contrasting colors on the floor.

Ceramic tiles in white and lime look great. Not necessarily stay with some sort of combinations. To make the kitchen bright and comfortable, it is better to use a white coating. It visually expands the space and uses a material of fine brown or black in the area of the chill-out.

Decorator "tricks" for beauty and comfort

If you place special decorations in the form of wall-mounted plates on the walls, the small kitchen will be cozy and beautiful. These details can be decorated with portraits of relatives and friends, landscapes that capture the soul, and even cute still-lifes.

On the eve of the new year, the image of the animal symbol of the next year will look very favorable. This applies to both textiles and small decorative items. The decision of such a decorator is especially typical of the Provencal style.

How to make cozy and natural materials

When the style of "A la French Village" takes into account that the entire fabric component of the design must be truly natural. Flax, cotton, mohair, and other materials that feel comfortable and do not hurt the eyes. How do you make a cozy kitchen? To give the room a festive atmosphere, you can use textile canvases in calm pastel colors. Accordingly, the furniture should look appropriate and not be knocked out of the general decoration.

A few words about zoning

For this purpose, the bar counter is perfectly suitable and solves both a practical and aesthetic problem. There, you can always hide the unsightly details that can "get the hell out of" every hostess.

Headset

How do you make a cozy kitchen? The best material for interior and interior design — an artificial stone, which is characterized by longevity, functionality, and ease of maintenance. With the help of this material, modern masters of interiors from ordinary spaces create true masterpieces of decorative art.

Let us note another advantage of artificial stone compared to natural stone. For an artificially created element, no great effort is required to lay it, while a brigade of professional masons is required for the alignment of this rock.

Prägnanz — the sister of the style

Speaking of many, imagine that *Naney cannot be a priori comfortable*. This is not the case, because beauty can be achieved not only by overloading the room with furniture but also by properly spaced nuances.

Usually, in such kitchens, the use of integrated technology is practiced, which is very convenient and compact. Some owners also put washing machines in their own kitchens hidden behind the closet doors. In kitchens, every inch of free space is often used minimally. The role of seats or tabletops in such kitchens can even be played through window sills with soft blankets and voluminous cushions. You just have to imagine how nice it is to look out the window in the evening, sit on such a window sill and enjoy a cup of tea or an aromatic coffee. By the way, about cups and other dishes.

In the style of minimalism, both perfectly white and bright dishes are greeted in rich colors. A set in such a kitchen can have any bright color (e.g., light green). Curtains on the windows (preferably white or with a modest imprint decoration, it is better if they are floral), given the question of "how to make a kitchen comfortable," it is very difficult to achieve real results it is necessary to choose the right textile zoning the room and decorating the room In order to achieve a completely unexpected result, the main focus is on the desire for change!

It can be all kinds of baskets, boxes, shelves,

compartments, and more. In many interiors, it is considered fashionable to provide the kitchen space with drawers that have a decent depth into which you can put all the necessary kitchen utensils and even large kitchen appliances.

All equipment for storing household items should be at arm's length so you can easily get what you need. The furnishing of furniture solves two very serious tasks at once. The first of these is the inner spatiality, and the second is the external compactness and the slight perception by the eye. In the space inside the table, all types of containers can be placed with different types of liquid and liquid products (butter, flour, cereals, etc.). The space near the sink is advisable to install useful units such as clothes dryers and other technical equipment. The area where the food is cut must be completely separate but at the same time, "parallel" to the main room of the room.

So we figured out how to make the kitchen cozy. Photos and practical tips in our article will help you solve this problem.

You will need the following:

- embedded household appliances,
- multifunctional household appliances,
- disk space,
- oven mitts,

- towels,
- tablecloth,
- curtains,
- picture,
- clock,
- family photos,
- board with colored pencils,
- magnets,
- children drawings,
- carpet,
- tile,
- multi-level lighting,
- TV mount,
- flowers
- instructions

Discard the door, or let it glide. You can also tear off the wall if it is not too worn, then create a studio. This extends the kitchen space.

Think through the space of a small kitchen and use it as sparingly as possible. Prefer the integrated technology and think about the storage system, which allows you to remove things that are rarely used from the public eye if the possible tailor-made kitchen. By measuring the space, you can use every corner of it. Do not think that small boxes are useless. You will appreciate them in practice when you realize that they can store many little things that every hostess needs, from knives to seasoning. Storage space can also be placed under the kitchen tabletop. You should not get carried away so you can sit comfortably. You can, however, place several small boxes under the tabletop. In a small kitchen, so-called corners can be conveniently placed. They allow much more economical use of the space, but at the same time, they offer the opportunity to place many people.

Discard artificial materials, buy wooden furniture. This gives a small room more comfortable. Glue wallpaper in the kitchen. Wallpapers create coziness, unlike a cold tile. Tiles should only be used to cover the "skirt" above the worktops, stove, and sink. Choose the warm colors of the material, avoid variations.

Make multi-level lighting, for example, consisting of the lighting above the work surfaces and the lampshade above the table. Hang a special stand on the wall where you can set up the TV. Arrange it so that it is convenient to see it when cooking but not while eating.

To create a cozy little kitchen, prefer dual purposes. For example, a built-in oven that also works in microwave mode saves space. It is not advisable to buy a fridge and a freezer separately. Choose a tall fridge with a built-in freezer with a large capacity. The surface above the built-in dishwasher and the washing machine can be used as a shelf or cutting table.

The biggest evil for small rooms, including small kitchens, is the crowded room. A small kitchen easily becomes a "warehouse" when the storage system is not well thought out, there are household appliances that are rarely used, and space is used uneconomically. Do not store anything in the kitchen. Do not store damaged items, lined dishes, and dark plastic containers. These things are not so expensive to affect the atmosphere of the kitchen.

An important role in creating comfort in a small kitchen is the decoration. It is worthwhile to set your taste. Collect the curtains, tablecloths, oven mitts, and towels that are combined with each other. In the kitchen, warm accessories are preferred, which are combined with the rest of the furniture. The pickup is worth it. To make a cozy little kitchen comfortable, hang a nice picture, a nice clock, or your family photos on the wall. Hang a board with crayons and children's drawings and attach some nice magnets to the fridge so you can take notes. That's how you make the kitchen individual, really yours. However, try not to overload the room with accessories.

So that the kitchen is not only delightful in terms of comfort and design, but also in terms of functionality and ergonomics, you must first eliminate unnecessary things. Then you have to fill the kitchen with necessary things, objects, and equipment that will facilitate the cooking process.

Rational Approach

Intelligent storage is primarily rationality. If everything you need exceeds the size of the kitchen, you will need to find new bins. They can fit even in the smallest spaces. You just have to see them.

In the smallest corner, for example, you can order a suitable rack size, for example, for spices, tea, and coffee. And with additional railing systems that are not just in the sink and oven area, you can place an infinite number of extra baskets where everything fits — from the bag for sandwiches to cat food.

If you do not have a built-in refrigerator, we recommend ordering a podium with a door underneath and a box with a shelf above it. Thus, there are immediately two additional memory slots to remove what needs to be removed.

On the windowsill, you can make a wide lining of artificial stone. It transforms a normal windowsill with a lonely, semi-dried geranium into a comfortable worktop that can be used to cook, as well as set up a teapot, coffee machine, toaster, blender, or scale.

Wherever possible, we recommend placing the shelves in different sizes and widths. They are suitable for storing banks, baskets, boxes, containers, and household appliances that are not used daily.

Additional storage space is provided by new stools, benches, and a kitchenette with internal drawers. All old chairs have to be replaced with it.

Clean Countertops

For many, all the surfaces of the plates and dining table look as if all the items had left the cabinets and were on strike. For cardinal changes in each of the things that stand, a suitable place must be found.

On the surfaces, nothing can be found except napkins, salt-shakers, a fruit bowl, and a few new books with recipes. Everything that is created in the cooking process should immediately return to its usual place.

After a few weeks of self-control, it becomes a habit.

If you use sliding racks with rotating mechanisms instead of the usual shelves behind the doors, then all the little things just fit in one place. In that case, you do not have to turn the entire cabinet to get it.

Always at Hand

What is used daily should be stored in the most appropriate kitchen? This is usually the place between the sink and the stove.

If there are too many necessary devices, add simple baskets to the cabinets that can be easily attached to existing shelves. In this zone, spices, daily cups, and glasses are best placed at the top of kitchen furniture.

In the lower cabinets and drawers, it is more logical to place daily dishes, cutlery, cutting knives, graters, a juicer, a meat grinder, combine harvester, and a blender. Now you can buy a variety of drawer dividers in home stores to separate the dishes from appliances.

To ensure that kitchen scissors, microscopic garlic graters, and spice mortars are not lost in the least possible time, you should attach hooks with Velcro to the inside of the cupboards in this area.

Spoons, spatulas, holders for paper towels, strainers, and drying cutlery can also be attached with rails to the wall between sink and oven. All of this will be available in an orderly manner.

Black Hole

Many in the kitchen storage tanks — a real black hole where everything is lost. Plastic organizers for storing and zoning help to correct the situation.

Holders for lids, containers for bulk goods and small parts, holders with suction cups that will not break the dishes, additional shelves, and mini-racks that increase the capacity of the cabinets — all this is very reasonably priced.

When planning storage, the simplest way is to divide all household items into multiple groups. It is such a plan that offers the opportunity to inform in advance about the locations for convenient use.

In the boxes, we recommend the use of several cutlery trays. In a prominent place are spoons, forks, knives, and underneath — not so often, hammers used for beating meat, rackets, working parts of the mixer, clamps for packaging, hose rolls for baking.

If you always lose the same items in your own kitchen, just put them in a separate container. It cannot only be kept in the drawer but also in a prominent place, especially if the container looks good and fits the overall style of the kitchen.

To make the kitchen comfortable, you should refuse to keep many things that may one day come in handy. All types of containers, bags, and containers are in the usable range. It makes more sense to create a functional, comfortable kitchen in the existing space.

To do this, you need to think about the location of home appliances, headsets, and other kitchen appliances. However, using the room properly is not the key to comfort. How do you create coziness in the kitchen so that space is functional and comfortable? To tackle this problem, you should be creative.

Minor Repair

First, the room should be refreshed. Eliminate unnecessary things that scatter the room. If the walls are covered with wallpaper, they should be changed. It is no secret that in the kitchen, all surfaces get dirty very quickly and become unsightly.

The tiling, if present, must be thoroughly cleaned of grease and dust, and the seams should be updated with a modern mortar. You can use a contrast that's combined with an updated interior design.

Create a cozy kitchen, and do not forget to arrange

the dining area. For decoration, you can use wallpapers with photo prints or paintings with bright colors. When you insert a colorful element into the arrangement of the room, you immediately see how the room has played with new colors.

Flooring and ceiling coverings are also worth updating in the course of the arrangement. For the ceiling, you can choose a shade of white. A floor should be a new laminate or linoleum. Such transformations do not require substantial material investment, and the visual effect will be mind-boggling.

Do not forget to bring the kitchen up-to-date. Old cabinets, shelves, and racks — it is necessary to clean the dust and grease, remove the old finish, and decorate. To do this, you can use the decoupage technique and stick to the set with napkins, photos, newspapers, and movie covers.

At the end of the work, the decorated surface must be painted to fix the decoupage and give the furniture a unique look. Furniture fittings must also be replaced by new interesting models. The old worktop covered with cracks and shavings should be thrown away. Instead, a new one should be ordered from modern, durable materials.

The Last Cuts

The room has already received new paint after its own work. Now a few details have to be added that creates a comfortable living kitchen. When decorating the room, use beautiful, original decorative objects. Do not fill the room with a large number of multicolored elements.

The most important design elements with which you can create a cozy atmosphere in the kitchen:

- Textiles — curtains, napkins, towels, tablecloths
- Kitchen utensils
- Light sources

Every homemaker knows that the curtains not only make the kitchen comfortable but can also change any room. For more variety, designers recommend two sets of curtains — for winter and summer. Window textiles should harmonize with the overall design of the room and give it a special boost. An excellent design step: kitchen textiles that are sewn from the same fabric as the curtains. It can be potholders, towels, decorative napkins, or chair covers.

A cozy kitchen can happen, perhaps, if beautiful

dishes are used.

Swap old cups, pots, and pans for new dishes and dishes. You will see how bright cups, plates, and pans literally revive the interior of the kitchen. Do not forget to add a candy vase, an original sugar bowl, and a container for spices to the place setting. These cute little things make for unique living comfort. This special mood allows you to create native, flowering plants in flower pots that can decorate the windowsill.

An important role in the design of a cozy kitchen in the photoplays the right lighting. Since the kitchen is a multifunctional space, several light sources must be used here. Be sure to highlight the workspace, racks, and shelves. In this case, headlights can be used.

In the dining area, the light should be soft and contribute to the absorption of food in a pleasant atmosphere. Near the table, you can use sconces or a modern sconce model.

The kitchen should have a central light that unites all functional zones in a holistic space. As a source of ceiling lighting, you can use a beautiful ceiling lamp that will illuminate the room with a bright, adequate light.

Creating comfort in the kitchen with your own

hands is, therefore, not difficult. Of course, you can strictly follow the advice of professionals and follow all instructions. In order to feel as comfortable as possible in the kitchen and feel the comfort and convenience, you should pay attention to your personal preferences. With your ideas and your creative work, you can set up a cozy and functional kitchen space.

16. CREATING A BALCONY: TIPS AND TRICKS

Whether you have a small balcony or large roof terrace: the design must be well-thought-out. First, it is necessary to consider whether the outdoor area should be used primarily for relaxation or whether the children should let off steam. If the latter is the case, one should take care when selecting the plants to ensure that they are robust and non-toxic. Try plants such as lilac, fuchsia, hibiscus, lavender, or ornamental fruit. Many people want to relax on the balcony and terrace, have a coffee with their neighbors, or have a barbecue with friends and family. With some easy-to-implement decoration ideas can make beautiful places to come together.

Railing: visual and wind protection

Tenants, in particular, have no influence on the condition of their balcony railings. In older homes, these are often not just an eye-catcher. The solution: balcony coverings made of weather- and UV-resistant textiles come in many colors, shapes, and patterns. They are easy to install, act as visual and wind protection, and are easy to remove when pulled out. Alternatively, reed or bamboo mats available in the hardware store fulfill the same purpose for the balcony border.

But it can also be stretched with a side awning or a fabric banner. Other space-saving options for privacy screens are wooden screens or movable walls, such as woven wicker or bamboo.

Those who prefer a natural green border can opt for climbing plants. Tightly planted together, they also hold the prying eyes of passers-by. Suitable for this are, for example, ivy, the black-eyed Susan, or the Dipladenia.

Flooring: carpets and wood instead of concrete

A gray concrete surface on the balcony is not for everyone. Balcony carpets are an alternative: these are made of special synthetic fibers that withstand sun and rain. In addition, the carpets are also easy to clean; some models are machine washable, while others only need cleaning with the vacuum cleaner. However, traditional carpets do not belong on the balcony, as they can fade in the sun or mold with moisture.

For those who do not like carpet outdoors, wooden tiles and plastic sheets are an easy and convenient option. Because the panels do not have to be glued first, they can be placed directly on the balcony floor and hooked together. However, care should be taken to ensure that there is still room for one centimeter between the new floor and the door.

Decorate the balcony with simple means: that's how it works

The final touch gets a balcony with a little decoration. It does not necessarily have to buy expensive decor items. For a change on the balcony, a few finds, such as stones or shells from the last holiday, shards of old flowerpots, or other decorative elements such as figures or balls care. These are simply laid next to the flowerpots on the ground, where they are gradually overgrown by the balcony plants. A balcony with self-designed terracotta pots is especially individual. The material is easy to paint with weatherproof paints and varnishes.

Tricks for a quick conversion: If you design your balcony with many different flowerpots and buckets, there is a decisive advantage. If visitors come or need more space for other reasons, the plants can simply be moved into a corner. For larger buckets, it is advisable to use small coasters with rollers, which are available in the hardware store or furniture store for little money.

Balcony and roof terrace in the right light
If you like to sit longer with a good glass of wine on the balcony in the evening, you can achieve great effects with atmospheric lighting. Tea lights can be placed in jam jars or vases filled with deco stones, shells, or sand.

Another idea: just punch holes in bags of breakfast bread paper and put the glasses and tea light in the bags. The flickering points of light create a special mood. Atmospheric lighting is also created by fairy lights that are attached to the railing or larger plants.

Especially in the city, many people have a small balcony of about two to six square meters: in the design of an even greater challenge, but not an impossibility.

Tips for Small Balconies

A small balcony clearly limits the space for creative development. But despite the lack of space, you can also do a lot with those square meters.

Plant-lovers who would like to accommodate

many plants on the balcony should not make the floor the sole storage area. If you attach the plants to the railing instead with different hanging devices, you can save a lot of space. Tomato shrubs on shelves, herb pots on hooks, or flowers in plant bags also visually bring a lot of variety. Another idea is special plastic flowerpots for the balcony edge: they consist of a plant bowl, which contains a recess in the middle so that they can be put on the wall or the railing.

If you like to cook with your own herbs, you can plant rosemary or basil on the balcony railing. They not only taste good, but they also look decorative. Sage or mint can also bring pleasant scents to a balcony. The herbs should, however, be housed in sufficiently large pots so that they thrive well. The pot you buy them in is usually too small.

Space-saving balcony furniture

For many people, the balcony is a place to meet with friends and family. For those who like to visit the small balcony, they need space-saving furniture.

Folding tables and chairs take up little space and can be stored quickly and easily if needed.

More space-saving options: a hanging table

can be easily attached to the balcony railing.

Side tables are smaller and lower than traditional tables and offer a shelf for glasses and cups. If you prefer to have a normal high table, you can opt for half a table. It looks like a table cut in the middle and can be placed close to the railing or on the wall. Not everyone wants to offer their guests a stool. A more elegant alternative to this are poufs: these are small seat cushions made of knitting, felt, or leather. But you should not be too exposed to the weather, so use weather-resistant polyester poufs.

Seats with a touch of extra are benches: with drawers below the seat or hinged seats, they can provide additional storage space. They can be placed on the narrow balcony side.

Sunbathers, especially those who go to the balcony to relax, need a suitable sunbed. On a small balcony, this should not be too massive; folding aluminum models are therefore well suited. And there is still the alternative to sunbeds: a hammock. Simply dowel two strong hooks into the walls and fix the hammock. Tenants should, however, before they intervene in the building of the house; as a precaution, ask the landlord for permission.

If the sun burns down too much, then sunscreen is important. However, conventional umbrella stands take away too much space with their massive base — one solution to this problem is using half-circle sunshades: these are placed with the straight side against the wall of the house. If the half-screen is not enough or if you want to avoid the overhang completely, a parasol-holder for the balcony railing is a better choice.

17. LIVING GUIDE: IT'S REALLY COZY

1. Design what you like most

A cozy living room starts with a good layout, because of the arrangement of furniture, carpets, lamps, and co. influences how well we feel in a room.

Go into the living room (ideally with the whole family) and think about how you spend your time there.

It is best to write down everything you would like to do there. This is so you will know what is important in the furniture arrangement.

Possible questions to ask yourself:
- Who "lives in" the living room? — family, friends, visit, playmates, pets...
- What do you most like to do in your living room? — watch TV, talk, read, play...
- What does everyone do by themselves? — And what do they all do together?
- Can you plan different areas for these activities?

Tip: Your living room is your home — not that of your guests. Even if you want to entertain (and impress) visitors to the living room, you and your wishes always come first.

2. Arrange furniture in groups

The denser the furniture, the more homey the atmosphere. Instead of a large furniture group, therefore, we often recommend small, separate units.

Depending on your favorite activities, such arrangements could be, for example:

- An area for chatting — with a comfortable seating area and a table
- An area around the game console, music system, and TV — with cozy cushions and armchairs to hang out on
- An area to play in — with soft carpet and room to run around
- A small reading area — with a large armchair, small table, and reading light

Tip: Amazing cuddly spaces are often off the living room couch — having many small sitting areas is fast becoming the favorite place for a cup of coffee or a

good book. A low table with chairs or a bench can provide a classy touch.

3. Determine areas with carpets

Carpets improve room acoustics and give furniture the optical hold.

And never to be underestimated: the cuddling factor.

When buying a carpet, it is important to find the right size. Because too large carpets can crush rooms, on the other hand, go under optically.

The following guidelines are helpful for the living room carpet:

Length of the carpet = length of the sofa plus 20 to 30 centimeters on the left and the right

About a third of all furniture, such as sofa and armchair, should have room on the carpets and rugs.

4. Pay attention to distances and paths

One of the most important tips for positioning your furniture:

You need to be able to move freely between the furnishing elements, and they cannot create any obstacles.

A good width for a main walkway in the living room is about one meter; for smaller ranges, a width of about 60 centimeters.

5. Pamper your living room with good light

Living rooms love good light.

And you will too, because light is hard to beat in terms of atmosphere.

Our most important tip: Never rely on one, but always on several lights.

Because no matter how beautiful it is, a single, bright ceiling light makes every room feel like a station hall. Coziness? Difficult.

On the other hand, islands of light consisting of one or more lights will divide the living room and give it depth:

- Floor lamps are great for smaller seating or reading corners
- Table lamps too
- Pendant lights can also be decentralized and slightly lower in the living room
- Low floor lights — for example, in a corner — are guarantors of coziness

Tip: On gray winter days, treat yourself to an extra dose of light — for example, with candlelight or a fairy light. Perfect.

6. Lay out blankets and pillows

Increase the fleece factor: Pillow. | Photo: Pexels

Coziness always has to do with warmth.
Think of a crackling fire or a cuddly blanket. But soft colors, friendly light, and soft materials also make rooms cozy.

Our next tip for the cozy living room is, therefore, to provide comfort and warmth with things that you like to look at and touch.

To start: fluffy blankets and pillows — for watching TV, reading, or sitting together.

7. Decorate them the most personally

Personal decorations are the most beautiful | Photo: Pexels

Coziness is also personality.
So that your living room does not look impersonal like from a catalog, confidently bring in your individual

touch.

What should you include?

- Things that you really love. Things that mean something to you. Things reminds you of dear people and beautiful events. They make you proud.

- Real photos of your loved ones — they always beat boring hardware store or furniture store posters, by a large amount.

- Personal souvenirs you brought back from vacation.

- Real art and homemade items — by people who really appreciate you.

You will see that personal accents make a room feel much more homey and interesting. They give you the feeling, "I am right where I belong."

And it gives your guests something to look at.

8. More courage with color

Many people give up their favorite colors in the apartment, because they worry it might look funny. Or worse, that others might find them funny.

However, colors can do so much: for example,

we can demonstrably calm down and radiate coziness, or we can be energetic and spray positive energy.

Colors can visually divide rooms and create niches — and make even white walls shine.

Our tip for a cozy living room: have the courage to include color.

Find the color concept that suits your needs and gives you positive energy and joy in life.

Finally, forget what others think. You live here.

9. Do not be afraid of your personal mix

As well as décor and colors, the interior may also be individually individual.

That means you do not have to surround yourself with furniture that fits together perfectly, because that would look boring and sterile.

Just have courage — and do not hesitate to put together furniture that you did not buy in the same furniture store. Old and new. Modern and Rustic. Wood and glass. Dots and stripes. It doesn't matter.

It is important that you feel well and that your living room suits you. Whatever pleases you is allowed.

10. Pay attention to the little things

Finally, a very special tip: Watch out for the little things, the things that signal to you and your loved ones that you are welcome.

Such little things often give a room more charm and coziness than a "real" sofa or the "perfect" color combination could ever do.

Put some fresh flowers out, put a few cookies in a bowl, make a pot of tea... and see how cozy your living room can look in a jiffy.

18. HOW TO CREATE COMFORT IN THE HOUSE

Family Comfort in Every Home

Comfort in the house is impossible to buy. You can invite a professional to create a unique interior, or you can buy a lot of attractive things for the house, but for some reason, the residents will not live happily in it.

There will be many similar interiors in which individuality is not felt.

Keeping in mind the lifestyle of the family and the habits of all family members, you should always strive to create a comfortable home. Friendliness, warmth, tranquility, comfort, comfort — those should be in it, and these qualities will only be present in a house where true love reigns.

What is meant by the term "comfort"? In all dictionaries, this concept is interpreted as a pleasant arrangement of life and personal environment. This comfort in your home can be created only by the owners themselves. No one knows the owners like they know themselves, and so they know that they need to get home as soon as possible after work and maintain that comfort.

Paradise is possible even in a hut. That's how it will always be in your hut:

- ruling individuality,
- mastering cleanliness and order,
- a lot of light and space,
- at least one houseplant, please, and
- always a fresh smell.

How do you create comfort in the house? We start with general cleaning.

Many women, as well as a portion of men, believe it is essential to promote a comfortable living space. Of course, even a simple, inexpensive renovation can refresh your home, filling it with light and adding visual space when all the surface materials are used brightly.

However, it is uncomfortable when, a few months after completion of the repair work, the handles and door knobs are covered with grease stains, a high-quality parquet looks like a floor in a barn, and new windows do not allow sunlight to pass through soiled or smudge-covered glass.

The order should always be there, and the comfort in the house will follow. If the host always tries to keep the house clean, even a bad environment is not apparent. What can be simpler than maintaining exemplary cleanliness in the house? If a homeowner does not work, it is.

And what about modern people who spend most of the day at work? You do not even have enough time every day to cook dinner and clean up the dust and vacuum your rooms.

Create a schedule in which a day in two weeks is dedicated to general cleaning. It is clear that it will be a day off and is best done on a Sunday. On Saturday, you can rest after the work week and communicate with friends and relatives, but the next day, you can begin to restore order with full force.

You cannot be fooled. If you remove the dust, do not leave a bit without your attention. If the items are on the countertops after ironing, place them so that they can be easily moved away later.

Keep summer items separate from winter items. When you clean the windows, do not forget to remove the bar in the corners of the window opening. Cleanliness should be everywhere, not just where it is visible. Spend a few hours (provided all family members help you: one parent, for example, washes the floor, while the children put away books and toys), and the house has already been rebuilt.

In a clean house, less chaos arises. Children who take part in general cleaning as much as possible create fewer disasters. All family members strive for comfort in the house. It may be advisable to schedule clean-up activities for each day of the week in the evening. A small clean-up that does not take a lot of time on an evening is time-consuming if all the items on the list are executed in one day.

Here is an example of how the work could be painted for a working person. They do not arrive home from work earlier than seven in the evening, but they always try to do everything they planned earlier:

- Monday — cleaning the toilet and bathtub
- Tuesday — cleaning the stove
- Wednesday — washing the kitchen and hallway floors
- Thursday — ironing
- Friday — vacuuming

Cozy Home with Your Own Hands: No Bad Smells

Another part of a cozy home: It should smell good. Nobody calls to bake cakes all the time so that the taste of the baking spreads to the landing. By pleasant odors, we understand above all a fresh smell.

In modern homes equipped with plastic windows, natural ventilation is disturbed: in winter, condensation forms on the windows and the corners are damped. Try to ventilate your home as often as possible.

If cleanliness is maintained, bad odors are less likely to occur. They are caused by dusty curtains, damp clothes/towels on the bathroom floor, outdoor shoes in the hallway, cat litter, and more. Do not leave dirty laundry in the washing machine, as this will create an unpleasant odor from the washing machine, as it is very difficult to remove.

Store food properly. There is nothing worse than the smell of rotten vegetables spreading throughout the house every time the fridge is opened — cleaning the refrigerator and checking everything in it should be done regularly.

A fresh smell in the house is good, and it can be supplemented with air fresheners. The main thing is not to overdo it, as no one likes a corrosive or overwhelming chemical smell. Make yourself at home with bags of fragrant, dried herbs, essential oils, scented candles.

Aromas of citrus, vanilla, cinnamon, bergamot — your sense of smell of such delicate notes will not be irritated but, rather, tuned to calm the nervous system.

House of warmth and comfort begins with the kitchen. Put a good dome on the soot and the smell of rancid oil will not go through the apartment. The presence of such high-quality kitchen appliances ensures that no grease deposits on the walls of the kitchen and all the furniture contained therein.

Create Comfort in the House with Indoor Plants

Indoor plants: incredibly simple, but always among the "top ten" of comfort ideas in the house. The harmonious combination of upholstered furniture with pots or other textile colors with the plant gives the possibility always to look stylish.

Each style means that there will be at least a small number of plants in the house. In a one-room apartment, you cannot accommodate much, but at least a few pots of useful indoor plants can be placed on the windowsill.

In the style of minimalism, when everything is extremely light, indoor plants become an accent — creating a bright habitat. Beautiful flowers look tall and are on the ground as a separate composition. Houseplants are not only purchased to provide comfort indoors but also to oxygenate the house. Do not forget that they absorb it at night, so not many plants should be in the bedrooms, especially if they are small.

Many live plants are preferable over artificial ones. Fake plants may look good, they require minimal care, but they are a real dust trap that cannot always thoroughly cleaned.

Buy a living flower, and it will...

- create comfort in your home,
- be a true healer for you,
- clean the air,
- improve the energy of the house,
- become a talisman, and
- help fruitful work.

It's just like setting up a young family to create coziness in their house with their own hands, and you do not have to spend a lot of money. Do not forget that your home should express your individuality; it should be a place where you can return to live in comfort and coziness and relax. Everyone wants to live in an ideal home, but it is impossible to get that without creating comfort.

1. Get Rid of Extra Things

Your house is neither a warehouse nor an abandoned property. So it's time to finally dismantle the cabinets and discard your accumulated junk/garbage in the name of comfort.

Disassemble clothes

Desperately spoiled and, frankly, old things a place in the dustbin, and those that seem decent but just do not like—you can try selling on Internet flea markets. Place your order in boxes that contained cosmetics and household chemicals. Everything spoils, so do not use or keep things you no longer want.

Same story with household appliances. A constantly overheating mixer, an iron with a battered cable, and wrapped tape seem to be trifles, but with frequent use, even small problems start to annoy.

At this stage, your goal is to get rid of everything that bothers you or gets on your nerves. A house is a place where we recover from fears and worries, so we should ensure that nothing bothers us here.

2. Add Unusual Details to The Interior.

A sterile, clean apartment where literally everything is straight and level is not for a *hyggelig* life. Small details help to get rid of the feeling of being in a hotel room.

Photos, paintings, an unusual tapestry, or a hand-tied napkin — such trifles lend personality. Your home is a reflection of your interests and your taste. So you can express yourself freely.

The only thing to remember is, above all, ensuring functionality. For example, a rug with a long nap might look nice near the sofa, but you'll regret buying it a thousand times every time you clean it. The same story with a wealth of statuettes and figures on the shelves — you must first remove them, then wipe away the dust and then put everything back in its place... Try to find a reasonable compromise in the choice between beauty and ease of cleaning.

3. Keep It Clean

In order not to kill for cleaning all weekend, divide it up in phases — it's easier than trying to cope with what has accumulated in a week in a day.

Here is a rough plan for each weekday. You can

add more tasks and change the order of their execution. The point is that maintaining cleanliness is a continuous process, not something you only do every two weeks.

- Monday: clean up the bathroom and arrange the laundry
- Tuesday: iron laundry.
- Wednesday: wipe off the dust
- Thursday: vacuum the floor
- Friday: wash the stove and remove any old food still in the fridge.

By the end of the week, everything is clean in your home. Only the little things like changing the sheets and washing the clothes you want to wear next week are left over. It's better to do the laundry on Saturday so you can iron it on Sunday and hang it up in the closet. The bed linen is changed on Sunday; on Monday, just do the laundry.

4. Simplify the cleaning process

The less effort required to restore order, the better. Prepare for the right inventory: with it, even general cleaning will not be a burden.

Wipe off the dust with microfiber cloths much faster and more efficiently than with old T-shirts or a piece of the old duvet cover. If you wash the floor with water with a special tool, you can wipe off the dirt the first time without much effort.

Hate sweeping floors? Buy a good vacuum cleaner, and you're done.

The Conditions Are Simple

It happens like that — you come to visit, and the house looks clean and light, and the style is understandable but uncomfortable. Or vice versa — it seems that there is nothing special in the house, but it is still cozy. Well, you do not want to go. Which comfort objects in the house create an atmosphere? What's up, is it just the hospitality of the hosts?

If the house is good
Luckily are those endowed with an inner sense of style and intuitively understand what a comfortable home should be. About creating comfort in the house, authored numerous articles, there are workshops, and lectures. The topic is still up to date. This book contains all sorts of tips and tricks for how you can create comfort in your home with your own hands.

Harmony in the room is created by adding simple but necessary things to the interior. You can literally find products for home comforts in every shop. For example, the usual *hyggelig* curtain can create the right atmosphere in the room, and if you do not follow simple conditions, only chaos is added.

When the rooms are small

If the rooms in the house are small, you should not overload them with heavy furniture. It is better to envision an economical but effective option where all lockers and shelves are strictly functional, and space is as meaningful as possible. There is no need to hang a heavy, multi-layered curtain in a small room; a simple, sweet curtain would suffice.

Flowers as comfort objects in the house

If a windowsill is available, then a small pot with a flower would be very appropriate. But if you cannot "get along" with fresh flowers, if they die and dry, you do not have to overpower and try to become an urgent florist. Dried, uncultivated plants only spoil the impression. In this case, go for small, artificial bouquets, especially as they can be placed not only on the windowsill but also on the upper levels of the cabinets.

If you're a remarkable florist, they come to you for advice on growing various household plants and seedlings. Then you know how to create coziness in your home and that flowers always create a positive mood. Plants planted in original flower pots, as well as artificial butterflies and ladybirds, give the freshness and beauty of the room.

Accordingly, if you are the owner of large rooms, you can choose any plant — even a palm in a tub or shy violets in small pots. In that case, all you have to do is select the place where the flowers look the best. If your rooms do not differ too greatly in size, you might prefer plants of medium and small size. And think of the numbers — too many flowering representatives of the flora can transform your home into an arboretum, and this will clearly bring no benefit.

Cozy windows

In addition to the colors that decorate the sills, you will need curtains to decorate the window space. The main condition for the selection of window curtains is respect for the general style of the apartment. For example, if Japanese minimalism prevails in your home, heavy multi-layer curtains look very inappropriate, just as simple paper blinds do not fit into a luxurious Oriental interior.

Lighting

The lighting in every room should be effective. For example, in the kitchen, a bright light is important, while in the bedroom, it should be soft and intimate.

Rooms that should be bright and brightly lit:

- the kitchen,
- bathroom and toilet, and
- the corridor.

For the bedroom and the dining room, soft, diffused light is preferable. In the living room, as well as in the children's rooms, it is better to have the possibility of changing the lighting from bright to soft.

Entrance hall

Where does a comfortable home start? That's right, from the hallway. It can even start earlier: from the front door and the doormat. Today, you can buy absolutely any rug: For inscriptions, pictures, and even 3D effects, it depends on your sense of humor and your financial resources.

There is nothing easier than to create comfort in the house when you start at the door. If the corridor allows it, set up a small sofa or soft puff cube so that you and your guests can wear and take off comfortable shoes.

Bedroom

As in every room of the house, you feel comfortable in the bedroom with small things. Incidentally, a multi-layered curtain looks good here, even if it visually hides the room because dusk and intimacy are necessary for this room. The bedroom is a space in which it is desirable to smooth the corners as much as possible. Even a straight-edged bed can soften if you cover it with a chic satin cover with a "soft" pattern.

In the bedroom, you can refuse the chandelier in the middle of the ceiling; a soft, diffused light is better suited here. Two table lamps on or

The kitchen

And here you need bright lighting at any time of day or night. A large window and some lamps around the kitchen are welcome; their concentration should be mainly on work surfaces.

The main rule of the kitchen is functionality. And yet you can implement different ideas for the comfort in the house here. All you have to do is to properly pick up the kitchen utensils in the same style and arrange the kitchen apron and the furniture. Rolls, elegant napkin holders, shelves of spices, cereals, and other things — these are the details and details of the comfort in the house.

Bathroom and toilet

In an effort to create harmony with the bathroom and toilet, many make the same mistake — they arrange different sweet knickknacks and hang many shelves. As a result, the opposite effect is produced, and the space appears crowded and cramped.

If you want your toilet to always look cozy, then remove all unnecessary items from there. Clear functionality and cleanliness are important in these rooms, and this is difficult to achieve when there are foreign objects. Especially focus on the appearance of various household items, buckets, brushes, and rags that are kept in the bathroom or on the toilet. For these items, it is better to choose a corner in the house, perhaps a wardrobe in the hallway. Or, if planning permits, make a *special Where* and remove all attributes for cleaning. If you still want to keep these items

226

in the bathroom, then make them a special locker, preferably in the basic colors of the room, as the main goal is to hide all unnecessary items.

In the bathroom itself, only two or three shelves should remain. Accents help to arrange the original towel-holders, doormats as well as cute plumbing stickers that you can choose according to your taste.

Living room

Here, you can let your imagination run wild and create comfort in your home with your own hands. Often, the living room is a recreational space for all households, a place for parties. Again, it is important not to overdo it and preserve the space — it should be pleasant enough for everyone. A soft, fluffy carpet, comfortable armchairs, cushions — this is an undeniably cozy home. The main thing: do not overdo it.

How to create comfort in a house where small children live? This is the most burning issue for young families with children of different ages. Even with their own room, children prefer to play where adults are, and so the living room often becomes a place where almost all of the baby's toys are unbalanced. Assign space for games, place a puzzle mat there, and place the basket for toys.

Fireplace

The fire in the fireplace always creates an atmosphere of solemnity and mystery, and the dim light only enhances the overall impression. The firewood burns, it crackles lightly, and the light of the fire falls on the lying, fluffy carpet, the cuddly, soft armchairs... romance.

If you cannot afford a real fireplace in your house, what will stop you from imitating them? Create the necessary space on one of the walls and get started! On the web you will find all sorts of tutorials to build a fireplace with normal cardboard boxes.

Take a lot of boxes of different sizes and fold a square sheet over them. Tape the box together with tape. Cover it with wallpaper with the image of bricks, and now fasten your fireplace to the wall — and voilá — a beautiful imitation of the mantelpiece is ready!

Decorate the room in front of the fireplace and dress it with a garland. It will sparkle with different lights and perfectly replace a natural source of fire. On the mantel, you can put pictures in frames, small stuffed animals, or arrangements of family porcelain.

Think about your family's crest, draw it, and attach it to the wall above the fireplace. With this, your house will find a special personality and a special charm.

For those who do not want to give up the real heat of a wood-burning fireplace without having the contraindications of a costly and complex installation and the production of toxic fumes during combustion, could opt for a modern solution that is economical, ecological and aesthetically extremely valid: a bioethanol fireplace. Bioethanol is an organic fuel extracted from the fermentation of biomass and, depending on the size of the chimney, an area of 20/30 sqm can be heated easily. The combustion produces water vapour and a modest amount of carbon dioxide, so it is ideal from an environmental point of view. As far as aesthetics are concerned, the models on sale have absolutely enchanting designs.

Or why not try to use your smart TV as a home fireplace by tuning it on Youtube and choosing one of those magnetic videos that transmit hours of dancing flames on lit logs as if you could really feel all the warmth and charm of a real fireplace?

System: "Fly Lady"

The main rule of a cozy home is cleanliness. Because even if famous designers have developed the furnishings of your house and famous construction companies have embodied the ideas, dirt, and dust in the house have destroyed the whole impression of the magnificent renovation, many homemakers have taken hold of the system "Fly Lady" for cleaning.

This form of cleaning takes exactly 15 minutes per day. After all, cleanliness and comfort in the home do not always require titanic efforts. The entire room of the house is traditionally divided into zones, and you will clean up one of the zones every day. As a result, the system will inevitably keep the house constantly clean, and you will not spend a whole weekend on the hassle.

In addition, there are many so-called life-hacking methods (and this is the secret of well-being in the house), which greatly facilitate the life of a housewife. For example, we all know that on the upper surfaces of kitchen cabinets constantly accumulates particularly greasy kitchen dust and dirt, whose removal is worth a lot of effort. If you cover the top of your lockers with a transparent cling film, you will see how much cleaning is reduced in this room. After all, you just have to replace the contaminated movie with a new one regularly. Cover the inside surfaces of kitchen cabinets to shorten the time it takes to clean the kitchen with the same foil.

Use an ammonia solution to glaze glass and mirror surfaces. Add some fabric softener to the water you want to remove the dust with. Then the dust will not bother you for a long time, and the house will smell sweet after cleaning.

But do not overdo it with purity, because the sterility of a museum ruins the charm of a house. If you hand guests boot covers and keep an eye on the things that touch them, you run the risk of not seeing anyone in your home, except for your own reflection in a perfectly clean mirror.

The atmosphere

Perhaps the most important rule of a cozy home is the atmosphere created not by furnishings or things, but by people. Cheerful, friendly hosts, happy guests, delicious food, laughter, lively conversations — these create the atmosphere of every home.

And do not forget the smell! Unpleasant odors will destroy all your efforts. So give them no chance to invade your apartment. All gaps in the walls, floor, and ceiling should be recognized and cemented to get rid of possessed cellar odors or amber coming from neighbors.

The kitchen of the cozy house is filled with the flavors of fresh cakes, coffee, and cinnamon, in the bedroom, and in the living room, there are light flower notes, and in the bathroom and toilet, there is a fresh aroma with notes of citrus or pine needles. Modern air fresheners cope with all these tasks.

In the kitchen, you can place a small bowl or bouquet made of coffee, cinnamon, and dried lemon, emitting a delicate, pleasant aroma. Flower petals cast in glass containers not only support the fragrant atmosphere but can also become a stylish part of the interior.

In other words, a cozy house is a house where people pay attention to little things, or rather, in such a house, there are no trifles. Everything is in its place, and everything is harmoniously in the room.

Many women want to create comfort in the house; it is nearly instinctive. Because of modern design, new, expensive furniture no longer makes a house comfortable. If you are concerned with repairs and thinking about decorating the interior with different accessories, they will give individuality and warmth. There are many things that you can do with your own hands, for example, sanding old furniture, painting a coffee table or dresser, or making a new lampshade. Ancient things will find a new life, and your home will be comfortable.

How to create comfort in the house

When creating a unique and comfortable interior, you must consider what the house is made of, how it is built, and what your individual preferences are.

It is much more comfortable to be in a house with a fireplace, even if it is an electric model. And you can look at the real fire and hear the crackle of the logs almost endlessly. Old wooden furniture can be sanded and covered with colored paint. Upholstered chairs, armchairs, and sofas are too tight and should be replaced with more comfortable seating.

Knitted napkins, tablecloths, embroidered pillows — all this in the interior will be used in the style of Provence or of the country. These beautiful little things will remind you of your relatives and give your home warmth and coziness.

How to make a room cozy

Choose the color scheme of the interior and

prioritize quiet, natural colors: light yellow, beige, brown, green, light blue. These colors have a soothing effect, filling the room with warmth and coziness. Trendy black and white, grays, and reds are recommended for offices, restaurants, and shops.

Some tips for designers to create comfort

For some, comfort in a room is associated with absolute order and cleanliness, while someone else may love "creative confusion" or "organized chaos" that gives a room a vivid look.

Personal preferences must be taken into account so that you feel comfortable and at home in your home. After all, a simple design will not add soul and warmth; it will make you sad.

Cozy home with your own hands

To make your home comfortable, it's important to get rid of all those old unnecessary things. Apart from the fact that they pollute the room, old things have bad energy. Clean regularly, and discard broken toys, label sheets, packaging, and boxes from purchases.

With various handicrafts, the interior can be given a piece of "soul." Embroidery, panels, mosaics, tapestries, and knitwear give a room a special touch. Embroidery and panels can be attached to the walls, and knitted carpets can adorn the floor.

You can also use wicker baskets and boxes for storage. Paper tubes are good for beginners, and you can use old paper, glue, and varnish. The flexibility of paper allows you to create unique things that you can store in clothing, textiles, and even in products.

If you have the sewing skills, you can sew beautiful curtains and bedspreads yourself. In this case, you can choose a fabric that is suitable for the interior and affordable. Soft curtains, pelmets, woven, or knitted lace can add elegance and comfort to the room.

Ideas for Comfort in Your House

The interior in every style can be decorated with a transparent round glass vase. You can fill it depending on the style of the room. You can install such a vase on a shelf or a coffee table.

Giving a new life to an old coffee table is possible with the help of the decoupage technique. The variety of designs allows you to choose the most suitable overall design. The old coating is removed with sandpaper; then, the decor is applied.

Knitted, woven, and sewn colored rugs not only

pamper your feet with warmth and softness, but they also add color accents.

Using state-of-the-art patchwork technology, we produce exclusive bedspreads, pillows, and night clothes from various pieces of fabric.

The quilling technique creates unique paper works that can be used to decorate bookshelves and cabinets.

To disassemble all the little things, books can be made into shelves to hold more books.

Deep down, the Hygge spirit is inside every one of us. It is enough to listen to that intimate voice that shows us where to find that satisfaction for our soul made of warmth, color, softness, cuddles, smiles and joy of sharing. Little beautiful things that make us more free and happy.